Nonlinear Wave Dynamics

Kluwer Texts in the Mathematical Sciences

VOLUME 17

A Graduate-Level Book Series

The titles published in this series are listed at the end of this volume.

Nonlinear Wave Dynamics

Complexity and Simplicity

by

Jüri Engelbrecht
Estonian Academy of Sciences,
Tallinn, Estonia

Springer-Science+Business Media, B.V.

A C.I.P. Catalogue record for this book is available from the Library of Congress

ISBN 978-90-481-4833-2 ISBN 978-94-015-8891-1 (eBook)
DOI 10.1007/978-94-015-8891-1

Printed on acid-free paper

Originally published by Kluwer Academic Publishers in 1997.

Softcover reprint of the hardcover 1st edition 1997

To my children Kersti and Mart

Table of Contents

Preface

At the end of the twentieth century, nonlinear dynamics turned out to be one of the most challenging and stimulating ideas. Notions like bifurcations, attractors, chaos, fractals, etc. have proved to be useful in explaining the world around us, be it natural or artificial. However, much of our everyday understanding is still based on linearity, i.e. on the additivity and the proportionality. The larger the excitation, the larger the response—this seems to be carved in a stone tablet. The real world is not always reacting this way and the additivity is simply lost. The most convenient way to describe such a phenomenon is to use a mathematical term—nonlinearity. The importance of this notion, i.e. the importance of being nonlinear is nowadays more and more accepted not only by the scientific community but also globally.

The recent success of nonlinear dynamics is heavily biased towards temporal characterization widely using nonlinear ordinary differential equations. Nonlinear spatio-temporal processes, i.e. nonlinear waves are seemingly much more complicated because they are described by nonlinear partial differential equations. The richness of the world may lead in this case to coherent structures like solitons, kinks, breathers, etc. which have been studied in detail. Their chaotic counterparts, however, are not so explicitly analysed yet. The wavebearing physical systems cover a wide range of phenomena involving physics, solid mechanics, hydrodynamics, biological structures, chemistry, etc. The experience obtained by analysing nonlinear waves shows that nonlinear waves and nonlinear oscillations are neighbouring facets of dynamics and have common roots. Still, nonlinear wave dynamics is just on the edge of describing chaotic phenomena and therefore needs close attention. This book hopes to add some ideas into this fascinating field.

The book does not follow the classical patterns of describing a physical problem from a very deep theory to practical applications. The nonlinear wave dynamics is shadowed by its complexity and therefore one should always think about possible special approaches in order to understand the physical phenomena and not to sink into mathematical difficulties. That is why this book is more like a personal view on nonlinear wave motion. Such an approach could certainly be argued—why a personal view? Could one present something absolutely new or stunning on deep and rather well-studied problems? This is a difficult question to answer, because frankly speaking every book has a personal touch, whatever the subject is. Even Bourbaki despite the plural behind the name has his/her/their own very special personal attitude. Here this personal touch is really specially stressed. Actually the idea is very simple and stems from Albert Einstein: "Everything should be made as simple as possible but not simpler." Another fruitful impulse came from books in physics reviewing problems from a certain unifying idea (self-similarity, symmetry, etc.). This has led to the style including so many questions as possible but leaving many questions without answer. Such an attitude was taken neither in order to irritate the reader nor because of the ignorance of the author. The reason is much simpler—we do not know all the answers yet. "The art of asking the right questions is more important than the art of solving them," said Georg Cantor, whose name is given to beautiful snowflakes. Who is going to decide whether the questions are right or not, is another matter.

The general idea advocated in this book is to start from complicated mathematical models describing wave motion as it results from the rules of continuum mechanics and then to find a simpler viewpoint that still keeps everything essential preserved. "One of the principal objects to theoretical research in any department of knowledge is to find the point of view from which the subject appears in its greatest simplicity," said J.W.Gibbs in 1881. After more than 100 years, this idea is still fresh. In wave motion, the simplest viewpoint is that of the one-dimensional, unidirectional wave. Exact factorization is possible only for the classical wave equation describing the "ideal" case of two waves. Every step towards reality involving dissipation, dispersion, nonlinearity, inhomogeneity, etc. makes the governing equation more complicated and the exact factorization is then evidently not possible. Every approximation should, however, be justified and estimated to keep what is considered essential. In many cases, these estimations are approximate and leave many ends open. Still, the model equations can be derived and they serve as standard bricks to measure complicated walls of science. In this sense, complexity and simplicity are just two facets of the same phenomenon, only we have to distinguish between then and be able to put the jig-saw puzzle together again, much against the warning of A.Toffler.

The general ideas mentioned above have certainly influenced the presentation in the book.

We start in Chapter 1 explaining some basic notions from wave dynamics. This is done for readers who are not familiar with the background. Chapter 2 presents some essential continuum mechanics with a special attention to the idea of internal variables. Cornerstones of nonlinear wave dynamics—the reasons for nonlinearity—are collected in Chapter 3. Several types of nonlinearities are analysed involving physical, geometrical, kinematical, and structural nonlinearities and also the nonlinearities of the combined type. Based on previous Chapters, the following Chapter 4 is devoted to the mathematical modelling of nonlinear wave phenomena. First, the general mathematical models are derived and second, the formalism of evolution equations is presented in order to reach explicit one-wave models. Several methods like the asymptotic (reductive perturbation), iterative, spectral, and other methods are described. In addition, a very important class—inertial-diffusive systems—is also briefly analysed that has the utmost significance when dealing with internal variables. Equipped with powerful methods of Chapter 4, it is possible to attack the complexities of nonlinear wave phenomena. Chapter 5 is devoted to complications in modelling of nonlinear waves. Space-dependent and time-dependent phenomena are described within the conventional theory of continua. Waves in complicated structures, however, need even more sophisticated approaches. The importance of being nonlinear is described for wave-guides and surface waves, and also for propagating instabilities. The latter phenomenon needs different mathematical models compared to the conventional evolution equations. Chapter 5 ends with the description of solitary waves paying the attention to the formation process of solitons generated by a harmonic input.

Chapter 6 collects selected case studies chosen for demonstrating the ideas of mathematical modelling given above. The first case—waves in layers—illuminates an unconventional way of deriving the evolution equations. The second case—waves with thermal effects—demonstrates how the choice of internal variables is related to the constraints of the process. The third case—waves in tissues—shows how many physical effects can be considered within one evolution equation. Last but not least, the fourth case—nerve pulses—is to demonstrate how the combination of observable and internal variables leads to a special second-order evolution equation.

Chapter 7 includes three essays of the author on nonlinearity, beauty, and complexity casting dynamics into a more general philosophical framework. The essays have been written earlier and two of them also published separately elsewhere (see acknowledgements). This is also a reason why some ideas and quotations are repeated in them. Every essay, however, is meant to be centered around a leading thread and may be read separately.

Chapter 8 is of the summarizing character stressing once again the importance of being nonlinear and discussing the future prospects in the field.

The structure of the book is consequently the following: simple basic arguments result in a complicated theory that needs again certain simplifications in order to understand the phenomena. Or in other words: the thread is from simplicity to complexity and from complexity to simplicity. Where possible, the verbal presentation is included beside the formulae in order to emphasize the essence of basic ideas. The material presented in the book is certainly selective in terms of the author's best judgement.

The author hopes that researchers and students (both undergraduate and graduate) interested in wave dynamics will find this book of interest. It does not mean that the field of readers should be only solid mechanics. The methods are rather general, especially the ideas of introducing diffusivity into the models. The case studies are drawn from examples that are of interest in seismology, thermomechanics, ultrasound in medicine, and nerve pulse transmission. It is expected that all those who are interested in the mathematical modelling of wave phenomena, may benefit from this book.

The book is essentially based on the previous research of the author. During the last decade, beside the research papers on this topic some reviews have been written and courses given (Helsinki University of Technology, 1991; RWTH, Aachen, 1992; CISM, Udine, 1994). Quite naturally, the book has outgrown from these reviews and courses. However, this has not been the only impulse. The author has been lucky to work in many European research centres—Cambridge, Paris, Aachen are just some examples—and, what is even more important—to have many colleagues as close friends. "No man is an iland intire of itselfe", said John Donne, and the imprint of the colleagues is undoubtedly in the book except for the shortcomings that are only due to the author. The discussions with Profs. G.I.Barenblatt, J.Ballmann, M.Braun, A.Jeffrey and G.A.Maugin have been especially stimulating. There are also the younger colleagues whose influence should be mentioned—this means a constant need for new problems and transparent explanation. The author owes thanks to all of them.

It is the real pleasure to thank Miss Pilvi Veeber, Mrs. Sirje Eskola and Mr. Ants Kivilo for their most valuable help in compiling the manuscript.

Grateful acknowledgement is made for the permission to reprint excerpts and figures from several sources:

– by permission of CISM (Udine): excerpts of Chapters 1–4 from J.Engelbrecht, Nonlinear Wave Motion and Evolution Equations, in A.Jeffrey and J.Engelbrecht (Eds.), Springer, Wien and New-York, 1994, 51–107;

– by permission of Estonian Academy Publishers: Sections 7.2 and 7.3 from Proc. Estonian Acad. Sci., Phys. Math., 1995, 44, 108–119 and 1993, 42, 107–118, respectively;

– by permission of Walter de Gruyter Publishers: excerpts of Section 6.4.4 from G.A.Maugin and J.Engelbrecht, A Thermodynamical Viewpoint on Nerve Pulse Dynamics, J. Non-Equilib. Thermodyn., 1994, 19, 9–23;

– by permission of J.Wiley & Sons, Inc.: excerpts of Section 5.2 from J.Engelbrecht, Waves, Solids, and Nonlinearities, Shock and Vibration, 1995, 2, N° 2, 173–190;

– by permission of Dover Publications Inc.: Fig. 1.8 (partly) from H.Kolsky, Stress Waves in Solids, 2nd ed., Dover, New-York, 1963;

– by permission of Cordon Art: Fig. 7.11, Birds Wallpaper of M.C.Escher.

Finally, thanks go to my constant supporter Kaja, my wife, who has patiently tolerated the long working hours of her husband.

Tallinn, December, 1996 J.Engelbrecht

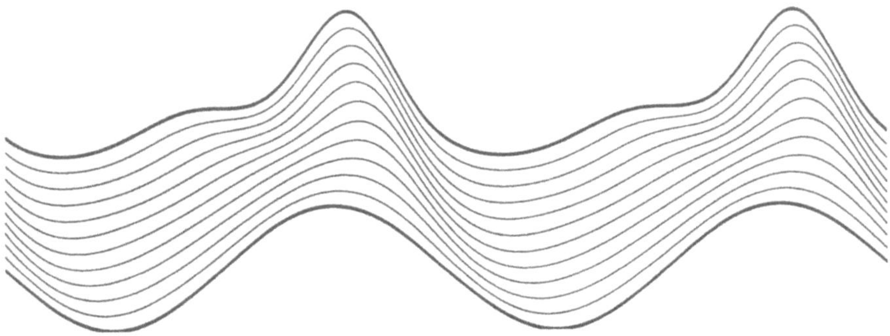

1. Introduction: basic wave theory

Theory is not solely an economical representation of experimental laws; it is also a classification of these laws.

P.Duhem

1.1. DEFINITIONS

Nature is full of waves. Waves in water and waves generated by earthquakes are probably the most familiar examples. During the last century, the notion "nonlinear waves" has been more and more exploited because wave phenomena turned to be much more complicated than predicted by conventional linear theories. The reason is not just in using high technology but also in our need to understand natural processes in their details. It is a huge task to write an overwhelming treatise about the wave phenomena. Here our aim is different—to understand the complexity of nonlinear wave motion. However, this topic is still too large and we have to restrict ourselves to certain specific problems. That is why we deal mostly with solids. This line is kept in order to demonstrate the background of physical considerations and mathematical models, the possible sources of nonlinearit-y/ies, the accuracy and limits of mathematical tools. Most of the examples are from solid mechanics but other examples are also included in order to emphasize some special features in wave phenomena.

The theory of mechanical waves in solids was developed during the 19th century by Cauchy, Poisson, Lamé and many others. For a long time it was acknowledged only as an extension of the theory of elasticity. The first reason for such an understanding was the level of the experimental technique that was not sufficient enough to measure stress waves and the second—the level of engineering was not high enough to use wave phenomena. In contemporary physics and engineering, waves in solids are easily generated and measured and the dynamical processes have got an important place in understanding nature and technology, and in measuring the various properties of materials and media. Even more, the need for accounting for properties of materials on the wide scale (microstructure, relaxation, anisotropy, etc.) and high intensity and/or short duration of excitations have led to much more complicated mathematical models than the conventional linear theory of elasticity that gave a starting impulse. These mathematical models must adequately describe complicated physical phenomena and along this line, nonlinear wave motion in solids has a special place in our physical world. This is not surprising because the world around us is essentially *nonlinear* and every theory developed to describe physical phenomena in their qualitative aspects, should take this basic property into account. The possible linearizations should be justified and used only in a certain range of parameters.

In order to build up an adequate theory or a mathematical model, the starting definitions should be clear.

First, what is solid? There are several definitions and here we give two of them:

Solid is a substance that has a definite volume and shape and resists forces that tend to alter its volume or shape.

Solid is a crystalline material in which the constituent atoms are arranged in a 3D lattice with certain symmetries.

The first definition is a basis for the theory of continuous media (for example, Eringen, 1962) and the second—for the theory of discrete media (for example, Kunin, 1982, 1983).

There is no accepted overwhelming definition of a wave because of its many facets. J.Pierce (1974) has said that better not ask what a wave is but ask what can be said about a wave. This way or another, here we would like to adopt the following definition (Truesdell and Noll, 1965):

Wave is a state moving into another state with a finite velocity.

A state in the above definition means stress, deformation, particle velocity, displacement or other measurable (observable) variables. We may also recall a wave as

a disturbance which propagates from one point in a medium to other points without giving the medium as a whole any permanent displacement.

In other words, a medium is deformed at a certain point and this disturbance is transmitted from one point to the next, etc. The resistance to deformation and the resistance to motion (i.e. inertia) must be overcome during the wave propagation. It is clear that waves can only occur in a medium in which energy can be stored in both kinetic and potential forms. Hence, the waves in solids are also characterized by the transport of energy through motions of particles that occur about an equilibrium.

Speaking about waves in solids, we first of all think about mechanical waves. However, in contemporary understanding some of the most interesting physical phenomena are nonmechanical (electric, magnetic, thermal, etc.) effects that accompany mechanical waves.

Two remarks are still needed. The first is about the importance of the causality principle: a wave can exist only due to an initial excitation or, in mathematical terms, due to boundary or initial conditions. The initial excitation is the starting point of the causal chain of interaction between perturbations and responses (energy transfer) and this can be realized only in finite time.

In addition, the second remark leads to an important facet of the causality. One has namely to distinguish between conservative and weakly nonconservative problems on one hand, and strongly nonconservative problems on the another hand. The first class of problems either neglects all energy in/outfluxes (linear theory of elasticity) or considers energy changes weak (linear theory of viscoelasticity). In this case the influence of the initial excitation is strong and traceable for the whole process. The second class of problems accounts for strong energy influx, coupled fields, etc. It may then happen that the influence of the initial excitation is weak and in the long run the wave characteristics are dictated by the properties of the medium. Here the initial excitation is needed to trigger the wave only and the causality principle is certainly not lost. A medium where such a process may occur is called active (see Engelbrecht, 1989).

1.2. TYPES OF WAVES

In order to start with clear definitions, we restrict ourselves to an isotropic elastic medium. It is easy to show (Kolsky, 1963; Bland, 1988) that the respective equations of motion correspond to two types of waves that can propagate in an unbounded elastic medium. These two types are characterized by comparing the particle motion with the direction of propagation:

If the particle motion is along the direction of propagation then the wave is longitudinal.

If the particle motion is perpendicular to the direction of propagation then the wave is transverse.

The schemes of these waves are depicted in Fig. 1.1a,b. Both of them are also frequently referred to as body waves. One should also notice that longitudinal waves are sometimes called as dilatational, irrotational, and extension waves, whilst transverse waves are called as shear, rotational, distortion, and equivoluminal waves. Within the framework of the linear theory of elasticity, these waves are uncoupled. According to more advanced theories this is not true and some interesting phenomena of polarization and coupling occur (Bland, 1969; Peipman et al., 1992). The velocity of a longitudinal wave in a certain medium is always larger than the velocity of transverse waves in the same medium. This was already known to Poisson, who gave the ratio of these velocities as $\sqrt{3} : 1$ (see Love, 1906).

If a solid has a free surface, then surface (Rayleigh) waves are possible. The particle motion in surface waves is in the plane perpendicular to the free surface and parallel to the direction of propagation (Fig. 1.1c). The amplitude of the surface waves decreases with the depth and their velocity is smaller than that of body waves. The same effect occurs along the boundary of two solids.

The situation is much more complicated in bounded media like rods, plates, shells, etc. This is caused by the multiple deformation states.

An example of possible deformation states in a cut-off of a plate is shown in Fig. 1.2 (c.f. Lee, 1976). The analysis of these waves forms an exciting chapter of elastodynamics (see Kolsky, 1963; Miklowitz, 1978; Bland, 1988).

One should also mention a very special type of waves that can exist in flexible

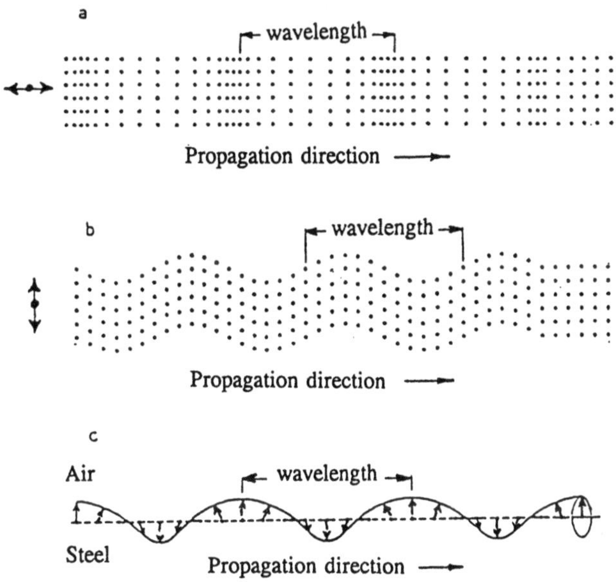

Fig. 1.1. Main types of waves: a – longitudinal, b – transverse, c – Rayleigh waves.

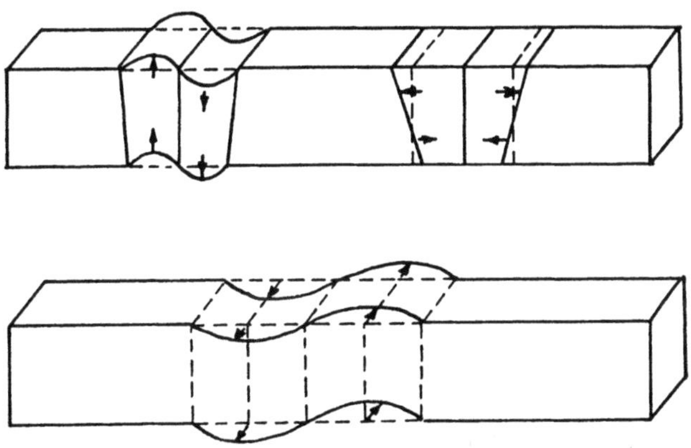

Fig. 1.2. Waves in a cut-off of a plate.

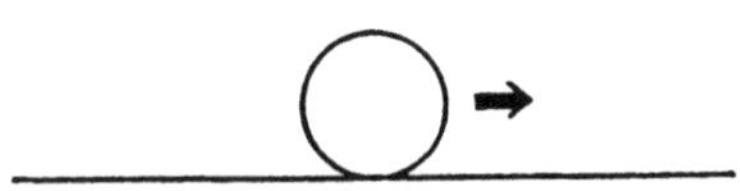

Fig. 1.3. Loop soliton in a flexible string.

strings (see Fig. 1.3). For obvious reason they are called loop solitons (Konno et al., 1981).

1.3. FAMILY OF WAVES

Intuitively speaking, a wave is characterized by its profile and its velocity. Here, in this subsection, a brief review on the wave profiles is given.

The first property is related to the smoothness of the profile: either smooth (in C^2 at least) or discontinuous. The sketches of these profiles are shown in Fig. 1.4. The second notion needs explanation—what is discontinuous. The governing wave equations have the leading terms (theory of elasticity) in the form of the second derivatives of displacement $U_i, i = 1,2,3$. A discontinuity is weak if it is in a derivative of order equal to or higher than the highest derivative of U_i (or other quantity) that appears in the governing equation. A discontinuity is strong if it is in a derivative of lower order. Supposing the displacement U_i be continuous (discontinuous U_i means a crack), the strong discontinuity is in $\partial U_i/\partial X_K$ or in the stress because the latter is related to the gradient of a displacement. Strong discontinuities are often referred to as shocks. Discontinuities in the second derivative of U_i are then weak—these are often referred to as acceleration waves (Truesdell and Noll, 1965).

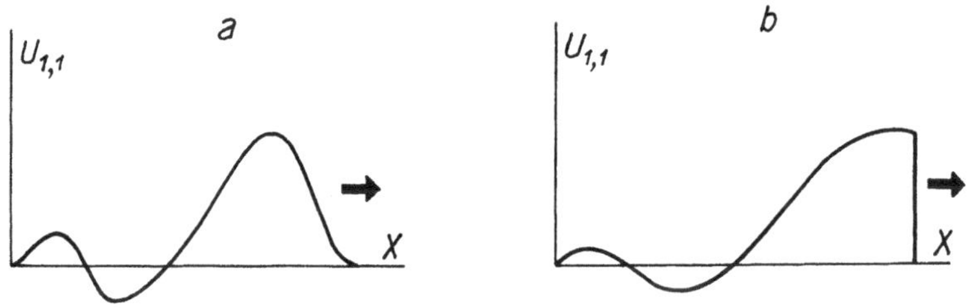

Fig. 1.4. Wave types: a – smooth profile; b – discontinuous profile.

According to the profile, the main wave types are of the impact type (Fig. 1.5a) and of the oscillating type (Fig. 1.5b,c,d).

In most cases the wave profile is dependent on time t and the propagation coordinate, i.e. there are at least two independent variables in governing equations (or 3 for 2D and 4 for 3D problems). Nevertheless, a possibility may exist that a wave profile is dependent only on one independent variable. In terms of t and $X_1 = X$, this independent variable is given like $\xi = c_0 t \pm X$, where c_0 is a velocity. Such a wave is usually called progressive (c.f. with the d'Alembert solution). The most familiar progressive wave is a harmonic wave that conventionally is described by the function

$$u = A \exp i\xi\,, \quad \xi = kX \pm \omega t\,, \tag{1.1}$$

where k is the wave number and ω the frequency. Here A may be the complex amplitude and k, ω are real.

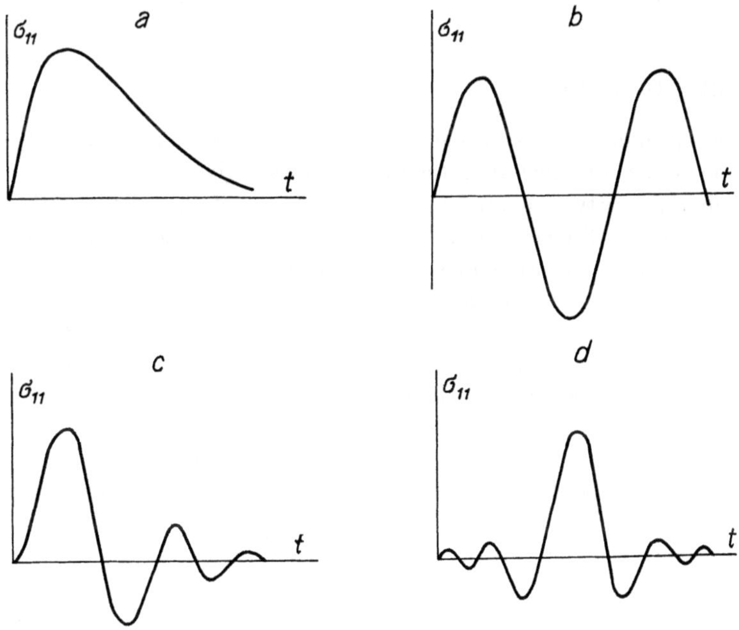

Fig. 1.5. Wave profiles: a – impact; b, c, d – oscillating.

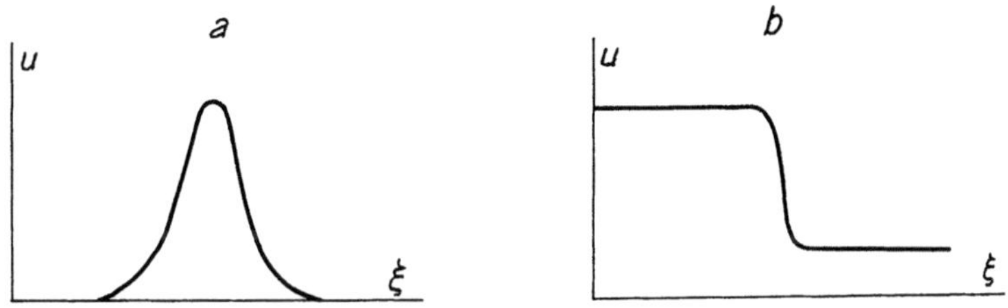

Fig. 1.6. Solitary waves: a – KdV soliton; b – kink wave.

In nonlinear theory, more waves may be of the progressive type. The celebrated family of solitary waves is of the utmost importance. The classical Korteweg-de Vries (KdV) soliton (Fig. 1.6a) and the kink wave (Fig. 1.6b) are the widely known examples. These waves exist due to a certain balance of nonlinear and other (dispersive, dissipative, etc.) effects. For more details, the reader is referred to Eilenberger (1981); Drazin and Johnson (1989), etc. In Section 5.3, the solitary waves are defined more accurately.

The solitary waves may also be asymmetric (Engelbrecht, 1991) with a certain overshoot as shown in Fig. 1.7 for a typical nerve pulse.

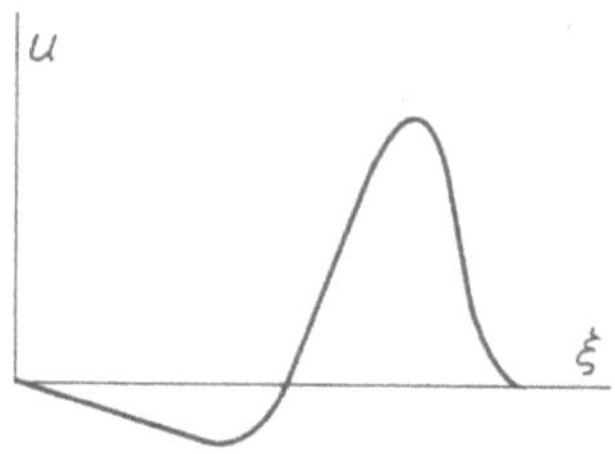

Fig. 1.7. A typical asymmetric solitary wave – a nerve pulse.

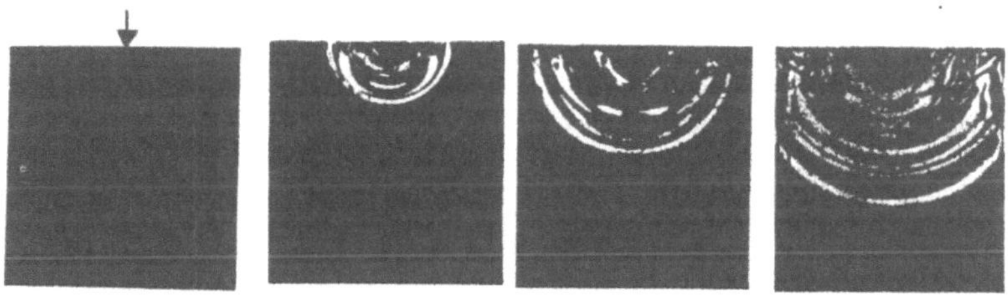

Fig. 1.8. Waves in a perspex plate (after Kolsky, 1963); from left to right: $t = 0$, 10.5, 21.7, 34.3 μsec, respectively. The arrow shows the place of excitation. Dover Publication, ©1963.

1.4. WAVE VELOCITIES AND DISPERSION

In Section 1.1, it was stressed that a wave must have a finite velocity. This is explicitly seen from Fig. 1.8 where the waves in a perspex plate due to an explosion at the upper edge are made visible by schlieren-technique (Kolsky, 1963). Here we shall briefly summarize the various concepts of wave velocities. As a prototype of the mathematical model, the simplest one-dimensional wave equation is used which is familiar from all the textbooks on wave dynamics. For a field variable $u = u(X, t)$ it reads

$$\frac{\partial^2 u}{\partial t^2} - c_0^2 \frac{\partial^2 u}{\partial X^2} = 0 . \tag{1.2}$$

The quantity c_0 is a constant and has the dimensions of velocity. Equation (1.2) is of the hyperbolic type and possesses the d'Alembert solution

$$u = f_1(X + c_0 t) + f_2(X - c_0 t) , \tag{1.3}$$

where the functions f_1, f_2 are to be determined from the initial and boundary conditions. The solution represents two waves, propagating in the positive and negative directions of X, respectively. The quantity c_0 is usually referred to as sound velocity. However, the real world is much more complicated than the model description by Eq. (1.2) and consequently the mathematical models are also more complicated retaining still the leading terms of the hyperbolic type as in Eq. (1.2).

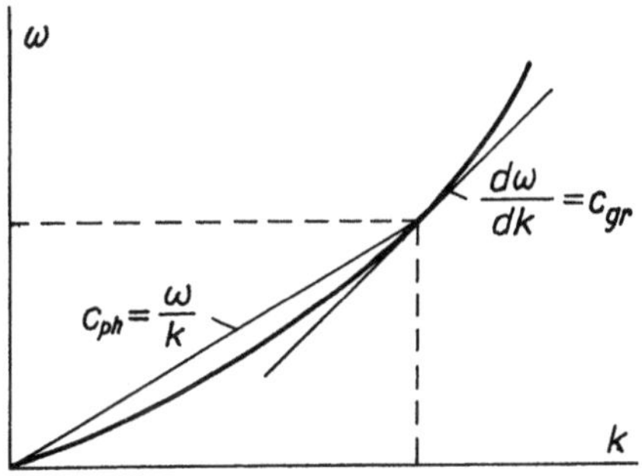

Fig. 1.9. Phase and group velocities.

It is easy to check that Eq. (1.2) is also satisfied by the harmonic wave solution (1.1) or its simplified variant

$$u = A\cos(kX + c_0 t)\,. \tag{1.4}$$

Rewriting this expression as

$$u = A\cos[k(X + \omega t/k)]\,, \tag{1.5}$$

it is easy to understand that the harmonic waves propagate with the velocity

$$c_{ph} = c_0 = \omega/k \tag{1.6}$$

that is called the phase velocity. The direct substitution of (1.5) into Eq. (1.2) gives the same result because

$$\omega^2 = c_0^2 k^2 \tag{1.7}$$

must be satisfied. In more general case, instead of (1.7) a more complicated expression

$$\omega = W(k) \tag{1.8}$$

is obtained with the frequency expressed as a function of the wave number. The phase velocity is then determined by

$$c_{ph} = \frac{\omega}{k} = \frac{W(k)}{k}\,. \tag{1.9}$$

If c_{ph} = const., then the wave is nondispersive, if $c_{ph} \neq$ const., then the wave is dispersive, i.e. waves with different wave numbers propagate with different velocities. Expression (1.8) is usually called a dispersion relation. For dispersive waves the group velocity, determined as

$$c_{gr} = \frac{d\omega}{dk} \tag{1.10}$$

is also of importance. The difference between phase and group velocities is shown in Fig. 1.9. In nondispersive case $c_{ph} = c_{gr} = c_0$, in dispersive case $c_{ph} \neq c_{gr}$. The energy transmitted by a wave propagates with the group velocity. For further details see Bland (1988).

Nonlinearity induces even more flavour into the family of velocities. The main point is that then the wave profile usually undergoes special evolution. This

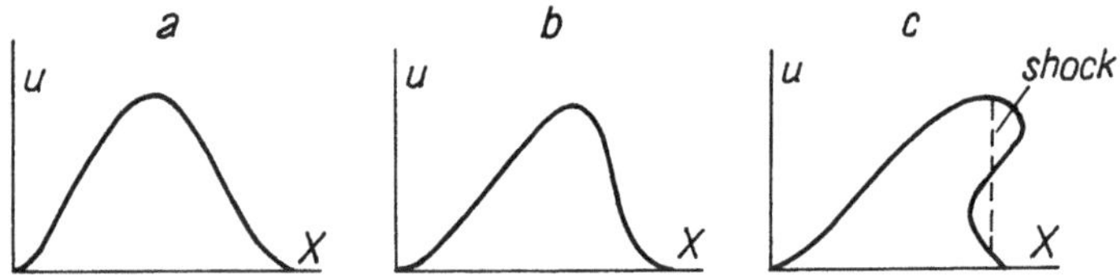

Fig. 1.10. Evolution of a nonlinear wave: a – $t = t_0$; b – $t = t_1 > t_0$; c – $t = t_2 > t_1$.

phenomenon will be demonstrated on a simple example. Let us imagine that in (1.2) sound velocity $c_0 \neq$ const. Instead of the 2nd order equation, we may also use its factorized simplification for one wave (Drazin and Johnson, 1989)

$$\frac{\partial u}{\partial t} + c(u)\frac{\partial u}{\partial X} = 0\,. \tag{1.11}$$

Let here $c(u) = c_0(1+u)$, for example. Then the solution to Eq. (1.11) is

$$u(x,t) = f(x - c_0(1+u)t)\,. \tag{1.12}$$

It means that the solution necessarily changes it shape as it propagates. The evolution is depicted in Fig. 1.10. At $t = t_2$, a shock is formed and its velocity c_{sh} is different from any of those velocities discussed above. Usually c_{sh} is larger than c_0, the sound velocity and depends upon the amplitude. For example, in one-dimensional case, considering only nonlinear elasticity, we obtain

$$c_{sh} = c_0\left\{1 + \tfrac{3}{4}(1+m_0)[U_{1,1}]\right\}, \tag{1.13}$$

where $[U_{1,1}]$ is the discontinuity of $U_{1,1}$ and m_0 is a certain nonlinear coefficient (Engelbrecht, 1983).

If there are more physical phenomena than the nonlinearity to be taken into account then the effects sometimes could be balanced and then the waves of constant profile can be observed. The solitary waves are one example of them, possessing again different velocities. The KdV soliton, for instance, propagates with the amplitude-dependent velocity c_s because of the balance of quadratic nonlinearity and cubic dispersion (see Section 5.3).

If viscous effects (fading memory) are taken into account along with the nonlinearity, then a steady wave may form with a certain velocity c_{st}. This velocity must be determined against equilibrium $c_e = c_0$ and instantaneous $c_i > c_e$ sound velocities (Nunziato et al., 1974). The latter results from the rate-dependent stress-strain relations.

In summary, this brief description of possible velocities in nonlinear wave motion should serve as a sign of attention. Details can be found elsewhere (Bland, 1969; Nunziato et al., 1974; Drazin and Johnson, 1989, and many others) but the reader must be aware of the importance of determining the correct values of velocities.

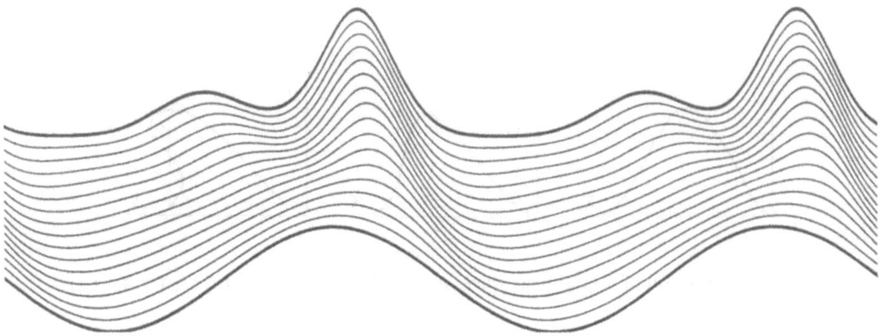

2. Essential continuum mechanics

We are to admit no more causes of natural things than such as are both true and sufficient to explain their appearances.
Rule of Parcimony, I.Newton

2.1. PRELIMINARIES

Continuum mechanics is concerned with the global behaviour of substances under external loads. Here we restrict ourselves mainly to solids, although some examples later cover a wider class of substances. The external loads act either through the surface of a body (contact loads) or by action at a distance (gravitation, for example).

"The rational mechanics of materials was begun by James Bernoulli, illustrated with brilliant examples by Euler, and lifted to generality by Cauchy" (Truesdell, 1984). This is a solid foundation indeed permitting to build a huge and complicated building of various theories and applications.

The classical nonlinear continuum mechanics is described by Truesdell and Noll (1965) and Eringen (1962), among the others. The most recent monographs are those of Smith (1993), Narasimhan (1993) and Chandrasekharaiah and Debnath (1994). For coupled fields, the monograph by Eringen and Maugin (1990) is of importance.

We give here a brief account of the general theory of continuous media. There are two definite subsections of the theory: (1) basic principles concerning conservation laws and thermodynamics and (2) constitutive theories. Both subsections will be briefly exposed further. However, the classical continuum mechanics deals only with observable variables like strain, temperature, electric field (if necessary), etc. Contemporary understanding (Maugin, 1990; Maugin and Muschik, 1994) underlines the importance of internal variables which are supposed to describe the internal structure of the body. The internal variables are independent and should be treated in constitutive theories with the same attention as conventional observable variables.

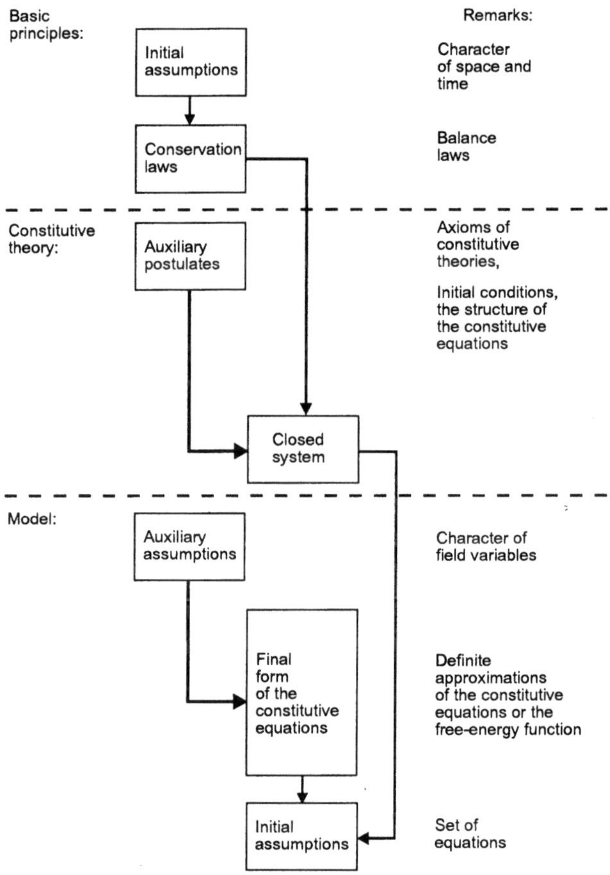

Fig. 2.1. Modelling architecture.

A more detailed account on internal variables is therefore included into this chapter (Section 2.4).

Our main aim throughout the book is to describe wave phenomena and continuum mechanics serves as a basis for constructing the mathematical models. A general scheme used for that is shown in Fig. 2.1. Only such a multi-level approach leads to a final mathematical model of an accuracy in which all the higher effects of the same order are taken into account.

Initial assumptions are the following:

(i) time—absolute time t;
(ii) space—three-dimensional Euclidean space;
(iii) macroscopic amount of substance B—continuous set of moving particles which at an arbitrary fixed instant, $t =$ const., exists in a space or region b which has surface A and volume $V \neq 0$ and possesses positive mass M, which is the sum of the masses of all parts of B.

Mathematical modelling of the behaviour of a macroscopic amount of substance B is possible using piecewise-continuous functions of the space (Euler) variables x^k, $k = 1,2,3$, or material (Lagrange) variables X^K, $K = 1,2,3$. Subsequently we fix upon the material description and use the rule of summation over the diagonally repeating index.

The laws governing motion are single-parameter transformations of coordinates:

$$x^k = x^k(X^1, X^2, X^3; t)\,, \tag{2.1}$$

$$X^K = X^K(x^1, x^2, x^3; t)\,. \tag{2.2}$$

Motion is called continuous if the right-hand sides of (2.1) and (2.2) are continuous functions and the transformations of coordinates (2.1) and (2.2) are mutually single-valued transformations.

2.2. CONSERVATION LAWS AND THERMODYNAMICS

Whatever the level of complexity of a process is, the conservation laws must always be obeyed. Actually they are also called balance laws and can be written either in the global or the local form. They consist of:

(i) conservation of mass;
(ii) balance of momentum;
(iii) balance of moment of momentum;
(iv) conservation of energy;
(v) entropy inequality.

In addition to stress and strain, the concepts of entropy, heat and temperature are inevitable in basic continuum mechanics. Said Eringen (1962): "Continuum mechanics is always darkened by thermal clouds." Our basic thread from complexity to simplicity is not aiming to account for or to sweep out the thermal clouds—one should just be aware of their existence.

We give here the balance laws in their local form because the constitutive theories follow then transparently. The following notations are used:

T^{KL} – Piola-Kirchhoff stress tensor;
E^{KL} – Green deformation tensor;
ρ_0, ρ – initial and current densities;
V, v – initial and current volumes;
f^k – the components of the body force;
A^k – the components of the acceleration;
$\left\{ {k \atop ml} \right\}$ – the Christoffel symbols;
E – internal energy;
Q^K – components of the heat;
h – the supply of the energy;
T – temperature;
S – entropy;
$F = E - TS$ – Helmholtz free energy.

The comma indicates the differentiation with respect to the coordinate and the dot—the differentiation with respect to time.

The verbal definitions are given by Eringen and Maugin (1990).

(i) conservation of mass: the total mass of a material body is unchanged during the motion of the body. It reads

$$\int_V \rho_0 dV = \int_v \rho dv \,. \tag{2.3}$$

(ii) balance of momentum: the time rate of change of momentum of the material body is equal to the resultant force acting upon the body. Actually, in an old-fashioned way and for a rigid body it may be rephrased like "the change of motion is proportional to the motive force impressed, and it takes place along the right line in which that force is impressed". This was stated by Isaac Newton in 1687.

In the local form we have

$$(T^{KL}x^k_{,L})_{,K} + \left(\left\{ \begin{matrix} k \\ ml \end{matrix} \right\} x^m_{,L}x^l_{,K} + \left\{ \begin{matrix} M \\ MK \end{matrix} \right\} x^k_{,L}\right) T^{KL} + \rho_0(f^k - A^k) = 0 \,. \tag{2.4}$$

(iii) balance of moment of momentum: the time rate of moment of momentum of the material body is equal to the resultant moment of all forces and the resultant of couples acting upon the body. This balance is of great importance for polar materials (see, for example, Eringen and Kafadar, 1976), for non-polar materials it states the symmetry of the stress tensor

$$T^{KL} = T^{LK} \,. \tag{2.5}$$

(iv) conservation of energy (the first principle of thermodynamics): the time rate of change of the sum of the internal and kinetic energies of a material body, considered as a closed system, is equal to the sum of the rate of work (or power) of all forces and couples and the energies that enter and leave the body per unit time. In the local form again we get:

$$\rho \dot{E} = T^{KL}\dot{E}_{KL} + Q^K_{,K} + \left\{ \begin{matrix} p \\ pl \end{matrix} \right\} x^l_{,K} Q^K + \rho_0 h \,. \tag{2.6}$$

(v) entropy inequality (the second principle of thermodynamics): the time rate of charge of the total entropy is never less than the sum of the entropy supply due to body sources and the entropy influx through the surface of the body.

Eliminating the sources and making use of the Helmholtz free energy, the entropy inequality is usually given as the Clausius-Duhem inequality:

$$T^{KL}\dot{E}_{KL} + \frac{1}{T} Q^K T_{,K} - \rho_0 \dot{F} - \rho_0 \dot{T} S \geq 0 \,. \tag{2.7}$$

Another form of this inequality in terms of the entropy flux $\mathbf{S}$ and the Cauchy stress σ is also used (see Section 2.4).

2.3. CONSTITUTIVE THEORIES

2.3.1. Auxiliary postulates and axioms

In order to get a closed system for determining the unknown field quantities, the constitutive equations must be formulated. According to the scheme shown in Fig. 2.1, the conservation laws must be complemented by auxiliary postulates and constitutive equation in order to get a constitutive theory. The conservation (balance) laws actually form the backbone of the theories but there is enough freedom at this stage to put some flesh around them.

Auxiliary postulates are:

1. the postulate of the existence of the initial state at $t = 0$,
2. the postulate of constitutive equations.

The latter determines the functional dependencies, not the exact form with definite coefficients. This final stage can be achieved only by knowing something quantitative about the process and will be analysed in Chapter 4.

The constitutive equations at this stage can also be presented verbally, like (Truesdell, 1984):

- the stress may be determined from the strain alone (perfectly elastic body);
- the stress may be determined from the stretching alone (perfectly plastic body);
- the flux of energy is a linear function of the temperature gradient (classical linear heat equation);
- the distances between particles do not change (rigid body).

It is clear that these statements actually include certain restrictions describing the phenomenological properties of materials. Consequently, these "constitutive equations" are valid only in certain domains of the variables involved. Clearly, some general physical and mathematical principles are needed to make order into the varieties of possible approaches.

These principles or requirements are also called axioms of constitutive theory (Eringen and Maugin, 1990). We follow the same pattern making some additional remarks.

The axioms are

(i) causality;
(ii) determinism;
(iii) equipresence;
(iv) objectivity;
(v) time reversal;
(vi) material invariance;
(vii) admissibility,

and two more that for more complex cases need special arguing:

(viii) neighbourhood;
(ix) memory.

For our purpose electromagnetic effects are excluded and for those the reader is referred to Eringen and Maugin (1990).

(i) axiom of causality:
The motion and temperature of the material points of a body are self-evident and observable in any thermomechanical behaviour of a body. The remaining quantities excluding the body force and energy supply that enter the balance laws and the entropy inequality, are the dependent variables.

(ii) axiom of determinism:
The value of Z at a material point $\mathbf{X}$ of the body B at time t, is determined by the history of all material points of B.

(iii) axiom of equipresence:
At the outset, all constitutive response functionals are to be considered to depend on the same list of constitutive variables, until the contrary is deduced.

(iv) axiom of objectivity:
The constitutive response functionals are form-invariant under arbitrary rigid motions of the spatial frame of reference and a constant shift of the origin of time.

(v) axiom of time reversal:
Thermodynamically this is equivalent to: the entropy production must be nonnegative under time reversal.

(vi) axiom of material invariance:
The constitutive response functionals must be form-invariant with respect to a group of transformations of the material frame of reference $\{\mathbf{X} \to \bar{\mathbf{X}}\}$ and "microscopic time reversal" as $\{t \to -t\}$ representing the material symmetry conditions. These transformations must leave the density at $(\mathbf{X}, t)$ unchanged. Here $\mathbf{X}, \bar{\mathbf{X}}$ are the neighbouring points.

(vii) axiom of admissibility:
Constitutive equations must be consistent with the balance laws and the entropy inequality.

There are now two axioms left that need some special attention. The point is how strong the axiom of determinism is, i.e. whether the material response is insensitive to the motion and temperature of distant points and to the history in the distant past. For "nonlocal" materials these axioms should be carefully analysed. However, in most cases the following axioms are of great help.

(viii) axiom of neighbourhood:
The values of response functionals at $\mathbf{X}$ are not affected appreciably by the values of the independent constitutive variables at distant points from $\mathbf{X}$.

This axiom can be reformulated in a more restrictive way as an axiom of local action (Truesdell and Noll, 1965). In determining the stress at a given particle X, the motion outside an arbitrary neighbourhood of X may be disregarded.

(ix) axiom of memory:
The values of constitutive variables, at a distant past from the present, do not affect appreciably the values of the constitutive response functionals at present.

This axiom can be made more restrictive by introducing the notion of fading memory (Coleman and Noll, 1961). It says: deformations that occurred in the distant past should have less influence in determining the present stress than those that occurred in the recent past.

Example 1: auxiliary postulates for the theory of viscoelasticity. These are formulated in the following form:

(1) Postulate of the existence of the initial state at $t = 0$:

$$\begin{aligned} &\dot{U}_k \equiv 0\,, \quad T^{KL} \equiv 0\,, \quad Q^K \equiv 0\,, \quad T \equiv T_0\,, \quad F \equiv F_0 \\ &T_0 = \text{const.}, \quad F_0 = \text{const.} \end{aligned} \tag{2.8}$$

(2) Postulate of the constitutive equations:

(i) The Helmholtz free energy is represented by functions of the form

$$F = F(I_1, I_2, I_3; T) \tag{2.9}$$

and has continuous partial derivations in I_1, I_2, I_3, T of all orders of magnitude which will be encountered in further consideration. Here I_M, $M = 1,2,3$, are the algebraic invariants of the deformation tensor E_{KL}.

(ii) The stress tensor contains the reversible $({}_ET^{KL})$ and the irreversible $({}_DT^{KL})$ parts; the components of the reversible part of the stress tensor ${}_ET^{KL}$ at each point of the material at a given instant t depend only upon the density of the Helmholtz energy at the same instant of time; the components of ${}_DT^{KL}$ at each point of the material at any given instant t depend only upon the deformation rate

$${}_DT^{KL} = {}_DT^{KL}(\dot{E}_{KL})\,. \tag{2.10}$$

(iii) The thermal conductivity equation has the form

$$\tau_0 \dot{Q}^A + Q^A = k^{AK} T_{,K} + \tau_0 \Gamma^{AKL} \dot{E}_{KL} \tag{2.11}$$

with $\tau_0, k^{AK}, \Gamma^{AKL}$ as material constants. Equation (2.11) is a modified Fourier law. A remark is here needed while (2.11) is based on the Cattaneo's equation which is not properly invariant under a change of frame. Despite of this shortcoming such modifications involving relaxation time τ_0 are widely used (Chandrasekharaiah, 1986). Müller (1987) has shown how to remove this discrepancy. Here Eq. (2.11) couples the fields of temperature and velocity as needed but requires further investigation with respect to its invariance properties.

From entropy conditions we get

$${}_ET^{KL} = \rho_0 \frac{\partial F}{\partial E_{KL}} - \tau_0 \Gamma^{AKL} T_{,A} T^{-1}\,. \tag{2.12}$$

For the irreversible part ${}_DT^{KL}$ we apply the Cauchy method (Eringen, 1962). Then the simplest approach gives

$${}_DT^{KL} = a_{01} \dot{E}_{KL} \tag{2.13}$$

where a_{01} is a scalar polynomial. The traditional approach of linear viscoelasticity gives (2.13) in the form

$${}_DT^{KL} = \left(\xi_\nu - \tfrac{2}{3}\eta_\nu\right) \dot{E}_{JJ} \delta^{KL} + 2\eta_\nu \dot{E}_{KL} \tag{2.14}$$

where δ^{KL} is the Kronecker symbol and ξ_ν, η_ν are the coefficients of viscosity.

Example 2: auxiliary postulates for the theory of thermal conductivity.

(1) Postulate of the existence of the initial state at $t = 0$ – same as (2.8).

(2) Postulate of the constitutive equations:

(i) The Helmholtz free energy is represented by

$$F = F(T)\,. \tag{2.15}$$

(ii) The thermal conductivity equation has the form

$$\tau_0 \dot{Q}^A + Q^A = k^{AK} T_{,K}\,. \tag{2.16}$$

2.3.2. Closed systems

Using the constitutive laws and the auxiliarity postulates we obtain the final closed system of the equations of thermoviscoelasticity:

$$
\begin{aligned}
&(T^{KL}x^k_{,L})_{,K} + \left(\begin{Bmatrix} k \\ ml \end{Bmatrix} x^m_{,L}x^l_{,K} + \begin{Bmatrix} M \\ MK \end{Bmatrix} x^k_{,L}\right) T^{KL} - \rho_0 A^k = 0\,,\\
&\rho_0 T\left(\frac{\partial^2 F}{\partial T \partial E_{KL}}\dot{E}_{KL} + \frac{\partial^2 F}{\partial T^2}\dot{T}\right) + {}_D T^{KL}\dot{E}_{KL} + Q^K_{,K} + \begin{Bmatrix} k \\ kl \end{Bmatrix} x^l_{,K}Q^K + \rho_0 h = 0,\\
&\tau_0 \dot{Q}^A + Q^A = k^{AK}T_{,K} + \tau_0\Gamma^{AKL}\dot{E}_{KL}\,,\\
&T^{KL} = {}_E T^{KL} + {}_D T^{KL}\,,\\
&F = F(I_1, I_2, I_3; T)\,,\\
&{}_E T^{KL} = \rho_0\frac{\partial F}{\partial E_{KL}} - \tau_0\Gamma^{AKL}T_{,A}T^{-1}\,,\\
&{}_D T^{KL} = a_{01}\dot{E}_{KL}\,.
\end{aligned} \tag{2.17}
$$

Simpler systems of equations follow from (2.17) if the auxiliarity postulates are changed. For example, the nonlinear constitutive theories of thermoelasticity, viscoelasticity, elasticity and thermal conductivity are easily deduced.

Several physical effects such as relaxation, creep and hereditary effects are described by means of integral operators. In this case, the histories of deformation and temperature are described by a special form of the free energy function (Christensen, 1971)

$$
\begin{aligned}
\rho F = \rho F_0 &+ \int_{-\infty}^{t} D^{KL}(t-\tau)\frac{\partial E_{KL}}{\partial \tau}d\tau + \int_{-\infty}^{t}\beta(t-\tau)\frac{\partial T(\tau)}{\partial \tau}d\tau\\
&+\frac{1}{2}\int_{-\infty}^{t}\int_{-\infty}^{t} G^{KLIJ}(t-\tau, t-\eta)\frac{\partial E_{KL}(\tau)}{\partial \tau}\frac{\partial E_{IJ}(\eta)}{\partial \eta}d\tau\, d\eta\\
&+\int_{-\infty}^{t}\int_{-\infty}^{t} \varphi^{KL}(t-\tau, t-\eta)\frac{\partial E_{KL}(\tau)}{\partial \tau}\frac{\partial T(\eta)}{\partial \eta}d\tau\, d\eta\\
&+\frac{1}{2}\int_{-\infty}^{t}\int_{-\infty}^{t} m(t-\tau, t-\eta)\frac{\partial T(\tau)}{\partial \tau}\frac{\partial T(\eta)}{\partial \eta}d\tau\, d\eta + \cdots\,.
\end{aligned} \tag{2.18}
$$

The functionals $D^{KL}, G^{KLIJ}, \varphi^{KL}, m, \beta$ determine the mechanical and the thermal properties of materials, being the continuous functions of their arguments and having a zero value for $\tau < 0, \eta < 0$.

The closed system takes now the following form (in rectangular coordinates):

$$
\begin{aligned}
&(T^{KL}x^k_{,L})_{,K} - \rho_0 A^k = 0\,,\\
&T_0\frac{\partial}{\partial t}\left[\int_{-\infty}^{t}\varphi^{KL}(t-\tau, 0)\frac{\partial E_{KL}}{\partial \tau}d\tau + \int_{-\infty}^{t} m(t-\tau, 0)\frac{\partial T}{\partial \tau}d\tau\right] + Q^K_{,K} + \rho_0 h = 0\,,\\
&Q^A = k^{AK}\tau_0^{-1}\frac{\partial}{\partial X^K}\int_0^t \exp\left(-\frac{t-\tau}{\tau_0}\right)T(X^K, \tau)d\tau\,,\\
&T^{KL} = G^{KLIJ}(0)E_{IJ} + \int_{-\infty}^{t} G^{KLIJ}\frac{\partial E_{IJ}}{\partial \tau}d\tau + \int_{-\infty}^{t}\varphi^{KL}\frac{\partial T}{\partial \tau}d\tau\,.
\end{aligned} \tag{2.19}
$$

The next step is to specify the constitutive equations on the basis of auxiliary assumptions. This will be done in Chapter 4 and before that we shall analyse nonlinearities that are inevitably accounted for in so many interesting cases.

2.4. INTERNAL VARIABLES

2.4.1. Idea of internal variables

As said before (Section 2.1), thermodynamics causes many complications in continuum mechanics. Briefly explained, the problem is how to describe the transition from one thermodynamic state to another under the restrictions imposed by the Clausius-Duhem inequality. The picture will be much clearer, i.e. the thermal clouds will be dispersed when the variables are distinctly separated into observable and internal ones. This approach is strongly advocated by Maugin and Muschik (1994) in their excellent review. Here we rely on these results and also on results of Maugin and Engelbrecht (1994).

First, a clear distinction between observable and internal variables should be made.

The *observable* variables are the usual field quantities like elastic strain and temperature, for example, that are observable in the real sense of the word. Moreover, they obey the balance laws (see Section 2.2) and are usually related to inertia. Observable variables are here collectively denoted by χ.

Internal variables, collectively denoted, say, by α are supposed to describe the internal structure of the body and are not observable. The examples could for instance be microscale structural rearrangements due to a dislocation movement, to the slip of grain boundaries, or to a phase transition induced by stress. Actually this means that internal variables should compensate for our lack of a precise description of a microstructure characteristic to polycrystalline media, periodic structures, etc. It might represent the average of some microscopic effect or the measure of some local structural rearrangement. This could be cases of viscosity, viscoplasticity, cyclic plasticity, rate-independent plasticity, damage, superconductivity, liquid crystals, etc. (see Maugin and Muschik, 1994). The internal variables used for the description of underlying microstructures can be real or complex valued scalars, vectors (polar or axial), and tensors.

Actually it is possible that for one process internal variables may have different character. For example, in polycrystalline plasticity a scalar variable is associated with the stress field arising from dislocations and point-defect clusters while a second-order tensor variable is related to the stress field due to cell walls or piles up of dislocations. The first variable is giving rise to isotropic hardening, the second one—to kinematical hardening (Bamman, 1985).

It is clear that a dependent variable, for example stress σ depends then simultaneously on both the independent observable and internal variables

$$\sigma = \sigma(\chi, \alpha) . \tag{2.20}$$

Consequently, a constitutive equation stated above must be complemented by a law which describes the temporal evolution of the variable α. For instance, we could write

$$\dot{\alpha} = f(\chi, \alpha) + g(\chi, \alpha)\dot{\chi} . \tag{2.21}$$

It is easily seen that $g(\chi, \alpha) = 0$ corresponds to the situation where the instantaneous strains are elastic (or zero).

However, whether a variable is an internal or observable variable, depends also on a researcher. Said Mandel (1980): “a clever physicist will always manage to detect the “internal” variables and measure them”. A variable could be considered internal, say, from the macroscopic observation point of view but on the other hand as observable from the mesoscopic viewpoint. Had a certain internal variable been considered as an additional observable variable, then a new balance law together with an inertia associated with this variable would have been needed. The governing equations for internal variables, as a rule, need a special experimental technique in order to establish functional dependencies.

Still, the choice between two natures of variables might be made by analysing the time scales at which the process related to both variables proceeds (Maugin, 1990).

2.4.2. Formalism of internal variables

It is convenient to use here not the balance laws but the canonical description involving the Lagrangian L. Such an approach facilitates introducing the dissipation potential $\mathcal{D}$ which is useful further on for deriving the definite mathematical models with internal variables.

The most important point is how to account for a possible dissipative mechanism related to internal variables in entropy production. According to Maugin (1990) it could be done by assuming

$$\mathbf{S} = \frac{1}{T}\mathbf{q} + \mathbf{k}, \tag{2.22}$$

where $\mathbf{S}$ is the entropy flux, $\mathbf{q}$ is the heat flux and $\mathbf{k}$ is an extra entropy flux density. Usually $\mathbf{k} = 0$ and (2.22) recalls the classical relation. Here we have to determine $\mathbf{k}$ from our analysis. We take $\rho_0 F = \psi$, denote the Cauchy stress tensor by σ and let the internal variable α be the n-vector (i.e. it has n independent components).

Now the Clausius-Duhem inequality (2.7) takes the form

$$-(\dot{\psi} + S\dot{T}) + p_{(i)} + \nabla \cdot (T\mathbf{k}) - (\mathbf{S} \cdot \nabla)T \geq 0, \tag{2.23}$$

where $p_{(i)}$ is the elementary power developed by the Cauchy stress (c.f. $T_{KL}\dot{E}_{KL}$ in (2.7)) and ∇ denotes, as usual, the gradient.

According to the scheme explained in Section 2.3 we have to state the structure of the Helmholtz free energy

$$\psi = \psi(\varepsilon^e, T, \alpha, \nabla\alpha), \tag{2.24}$$

where ε^e is the elastic strain tensor ($\varepsilon = \varepsilon^e + \varepsilon^p = (\nabla\mathbf{u})_s$, while ε is the total strain in terms of the displacement $\mathbf{u}$ and ε^p is the anelastic strain; we consider small strains only). This relation yields

$$\dot{\psi} = -S\dot{T} + \sigma : \dot{\varepsilon}^e - \mathcal{A}\dot{\alpha} + \nabla \cdot (\mathbf{B}\dot{\alpha}), \tag{2.25}$$

where the following notations have been used:

$$S = -\partial\psi/\partial T, \quad \sigma = \partial\psi/\partial\varepsilon^e, \tag{2.26a}$$

$$A = -\partial\psi/\partial\alpha, \quad \mathbf{B} = -\partial\psi/\partial(\nabla\alpha), \tag{2.26b}$$

$$\mathcal{A} = A - \nabla \cdot \mathbf{B} = -\delta\psi/\delta\alpha, \tag{2.26c}$$

$$\frac{\delta}{\delta\alpha} = \frac{\partial}{\partial\alpha} - \nabla \cdot \frac{\partial}{\partial(\nabla\alpha)} \,. \tag{2.26d}$$

The extra entropy flux is selected as

$$\mathbf{k} = \frac{1}{T}\mathbf{B}\dot{\alpha} \,, \tag{2.27}$$

and the residual dissipation inequality follows in the form

$$\phi \equiv \sigma : \dot{\varepsilon}^p + \mathcal{A}\dot{\alpha} - (\mathbf{S}\cdot\nabla)0 \geq 0 \,. \tag{2.28}$$

Often this inequality is split into two parts (i.e., stronger conditions are taken)

$$\phi_{\text{intr}} = \sigma : \dot{\varepsilon}^p + \mathcal{A}\dot{\alpha} \geq 0 \,, \tag{2.29}$$

$$\phi_{\text{th}} = -(\mathbf{S}\cdot\nabla)\theta \geq 0 \,. \tag{2.30}$$

Here ϕ_{intr} and ϕ_{th} indicate intrinsic and thermal dissipation, respectively.

Thermodynamical equilibrium is defined by the condition of spatially uniform temperature and the two conditions

$$\dot{\varepsilon}^p = 0 \,, \quad \mathcal{A} = 0 \,. \tag{2.31}$$

We postulate the existence of a dissipation potential

$$\mathcal{D} = \mathcal{D}(\sigma, \dot{\alpha}; \varepsilon^e, T, \alpha, \nabla\alpha) \tag{2.32}$$

which is a positive function, homogeneous of a definite degree in σ and $\dot{\alpha}$, and convex in these two variables so that

$$\dot{\varepsilon}^p = \partial\mathcal{D}/\partial\sigma \,, \quad \mathcal{A} = \partial\mathcal{D}/\partial\dot{\alpha} \tag{2.33}$$

exist. Now, on account of (2.25) and (2.26c), we record

$$\frac{\delta\psi}{\delta\alpha} + \frac{\partial\mathcal{D}}{\partial\dot{\alpha}} = 0 \tag{2.34}$$

that must be read together with (2.24) and (2.32).

The dissipative nature of (2.31) directly stems from the manner in which it was obtained and, obviously, from the fact that α is considered as an internal variable. We may introduce the Lagrangian per unit volume

$$\mathcal{L} = \mathcal{K} - \psi \,, \tag{2.35}$$

where $\mathcal{K}$ is the kinetic energy. Then (2.34) reads

$$\frac{\delta\mathcal{L}}{\delta\alpha} - \frac{\partial\mathcal{D}}{\partial\dot{\alpha}} = 0 \,, \tag{2.36a}$$

$$\frac{\delta}{\delta\alpha} = \frac{\delta}{\delta\alpha} - \frac{\partial}{\partial t}\frac{\partial}{\partial\dot{\alpha}} \,. \tag{2.36b}$$

If now the relaxation effects dominate over inertial ones, then (2.36) yields

$$\frac{\delta\psi}{\delta\alpha} + \frac{\partial\mathcal{D}}{\partial\dot{\alpha}} + \mu\frac{\partial}{\partial t}\left(\frac{\partial\mathcal{K}}{\partial\dot{\alpha}}\right) = 0 \,. \tag{2.37}$$

A scaling factor μ (a small parameter) denotes that there are inertial effects associated with α which will be discarded. In the first approximation we get

$$\frac{\delta\psi}{\delta\alpha} + \frac{\delta\mathcal{D}}{\partial\dot{\alpha}} = 0 \,. \tag{2.38}$$

On the contrary, if dissipation produces only a perturbation, (2.36) may be rewritten as

$$\frac{\partial}{\partial t}\left(\frac{\partial\mathcal{K}}{\partial\dot{\alpha}}\right) + \frac{\delta\psi}{\delta\alpha} + \mu\frac{\partial\mathcal{D}}{\partial\dot{\alpha}} = 0 \,, \tag{2.39}$$

where the scaling factor emphasizes the dissipative contribution. Again in the first approximation, (2.39) yields

$$\frac{\partial}{\partial t}\left(\frac{\partial \mathcal{K}}{\partial \dot{\alpha}}\right) + \frac{\delta \psi}{\delta \alpha} = 0 . \tag{2.40}$$

The two equations (2.38) and (2.40), both deduced from (2.36) show explicitly the qualitative differences between them. The question is whether the dissipative effects dominate over inertial ones or vice versa. If dissipation is small (μ is small) then we obtain the governing equation (2.40) which describes wave-like phenomena with inertia (kinetic energy is involved). Equation (2.40) could also have been derived from balance laws. With strong dissipation, α is a real internal variable and its governing equation (2.38) is of the evolution-diffusion type (inertia is discarded in the first approximation).

To sum up, we have now considerably enlarged our modelling rules. Introducing the concept of internal variables, we also must enlarge our postulates. In Section 2.3, the postulates were given for observable variables only. If we have both the observable and internal variables to be accounted for, then the existence of a dissipation potential should also be postulated.

The list of postulates is then:

(i) postulate of the existence of the initial state at $t = 0$,
(ii) postulate of the constitutive laws for observable variables;
(iii) postulate of the dissipation potential.

Several examples of using the formalism of internal variables are analysed in detail by Maugin and Muschik (1994) with a long list of references. In this treatise, we shall use the concept of internal variables when explaining the modelling of waves (Chapter 4) and analysing the case studies (Chapter 5).

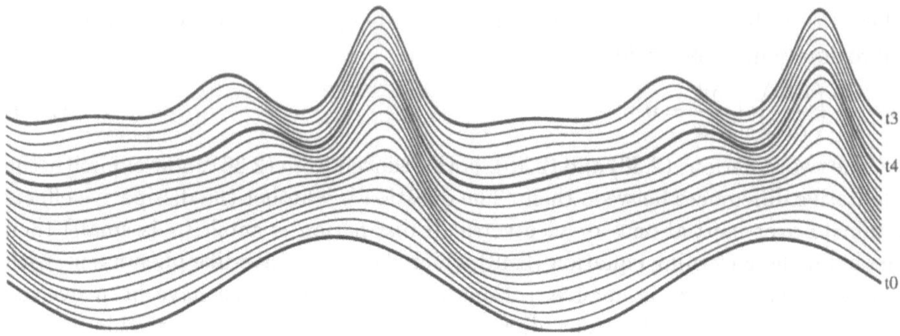

3. Nonlinearities: cornerstones for complexity

> *... the straight line leads to the downfall of mankind. ... it is the line which does not exist in nature.*
>
> *F.Hundertwasser*

3.1. BASIC IDEAS

Nonlinearity, non-linearité, Nichtlinearität, non linearità, mittelineaarsus (Estonian), . . . —what a sequence of no's in so many languages! This peculiar notion with the negative prefix needs some explanation. Starting from the word nonlinearity itself, it reflects our historical understanding of Nature — "linear" was primary in all the physical and mathematical theories invented by mankind. Contemporary science may in this sense be traced back to Leonardo da Vinci who suggested the following (Truesdell, 1968):

1. Observe the phenomenon and list the quantities having numerical magnitude that seems to influence it.

2. Set up such linear relations among pairs of these quantities as are not obviously contradicted by experience.

3. Propose these "rules of three" for trial by experiment.

The famous Second Law of Newton has put a seal to this understanding because "the change of motion is proportional to the motive force impressed", as stated in Principia, 1687. Since then the generations of scientists have been trained to think "linearly". Although here and there nonlinearity had a certain meaning, the real breakthrough came only about the middle of the 20th century. Nowadays the importance of being nonlinear (West, 1985) is quite well understood because, generally speaking, most of the physical (chemical, biological, etc.) processes do not possess the property of proportionality and the property of independence. In mathematical sense, emphasized either algebraically or geometrically, these properties have different implications. Table 3.1 summarizes some general understandings how the linear and nonlinear processes differ.

TABLE 3.1.

Excitation	Process	Result
F_1	linear	r_1
F_2		r_2
αF_1		αr_1
$F_1 + F_2$		$r = r_1 + r_2$
F_1	nonlinear	R_1
F_2		R_2
αF_1		$R_{\alpha 1} \neq \alpha R_1$
$F_1 + F_2$		$R \neq R_1 + R_2$

Briefly said, the adjective "nonlinear" means using a mathematical notion illustrated in Table 3.1 for describing a very special quality—an imprint of nature forbidding additivity in many aspects. In this sense we may recall L.Carrol who has let Humpty Dumpty to answer Alice arguing about making the words mean different things: "The question is, which is to be master—that's all."

Nonlinear phenomena play an increasing role in many areas of science—condensed matter physics, hydrodynamics, solid mechanics, acoustics, chemistry, biology, etc. In this sense, one may speak of nonlinear science that is essentially interdisciplinary because of the vast majority of natural phenomena and their mathematical models are nonlinear while linearity is an exceptional but instructive case. Nonlinear dynamics with its numerous paradigms such like solitons, strange attractors, fractals and many others, has drastically changed our views not only because of discovering new effects but also because of changing the cognitive power of mankind. This means first of all wider understanding about the possibilities to predict the course of the process or events.

In the previous Chapter the constitutive equations in solid mechanics were briefly presented. Further, the possible sources of nonlinearities are analysed and we try to take a broader view, not restricting ourselves only to stress and deformation waves in solids. The point is in the universality of many basic notions of dynamics, characteristic to many fields. Still in this Chapter the physical viewpoint is kept while in Chapter 7 this important notion is analysed from the viewpoint of cognition.

3.2. PHYSICAL NONLINEARITIES

The passage of a wave may be interpreted as the perturbation of an initial equilibrium state. It means that the constituent parts of a body act on each other causing contact forces. The extrinsic contact force per unit area gives rise to the stress hypothesis. Stress is one of the main components in conservation laws (c.f. Section 2.3) and is manifested by an expression relating stress to strain, temperature and other observable and/or internal variables. The nonlinearity of this expression is usually called physical or material nonlinearity.

According to the conventional continuum theory (Eringen, 1962), the stress tensor is related to the internal energy satisfying the entropy conditions. Introducing the Helmholtz free energy F and following the Green's method, the Piola-Kirchhoff stress tensor is determined as

$$T^{KL} = \rho_0 \frac{\partial F}{\partial E_{KL}} \tag{3.1}$$

and another pseudostress tensor as

$$T_l^K = \rho_0 \frac{\partial F}{\partial x^l_{,K}} . \tag{3.2}$$

If only rectangular coordinates are used then a simpler variant with lower indices only gives (Bland, 1969)

$$K_{ij} = \rho_0 \frac{\partial F}{\partial E_{ij}} , \tag{3.3}$$

$$L_{ij} = \rho_0 \frac{\partial F}{\partial U_{i,j}} , \tag{3.4}$$

$$L_{ij} = K_{kj} x_{i,k} . \tag{3.5}$$

Here K_{ij} is known as the Kirchhoff stress tensor and L_{ij} as the Lagrangian stress tensor.

Hence, the nonlinearity of the stress-strain relation depends upon the structure of the free energy. It is useful to remember the basic properties of the continuum. The potential energy U *vrs* the distance r between the atoms of a solid is schematically depicted in Fig. 3.1. and the curve is highly nonlinear indeed. From (3.1) a.o. it follows that in order to get a linear stress-strain relation, the energy can not contain higher than the second order terms in E_{KL} (or E_{ij}), i.e. the energy curve should be parabolic that is not the case. Only for small deformation (changes in r) around the minimum such a parabolic approximation holds with a sufficient accuracy.

In order to satisfy the invariance requirements, the free energy is formulated in terms of three independent algebraic invariants of the strain tensor

$$I_1 = E_{ii} , \quad I_2 = E_{ij} E_{ij} , \quad I_3 = E_{ij} E_{ik} E_{kj} . \tag{3.6}$$

One possible free energy function (potential) can be approximated as follows

$$\begin{aligned} \rho_0 F = {} & \tfrac{1}{2} \lambda I_1^2 + \mu I_2 + \nu_1 I_1^3 + \nu_2 I_1 I_2 + \nu_3 I_3 \\ & + \kappa_1 I_1^4 + \kappa_2 I_1^2 I_2 + \kappa_3 I_1 I_3 + \kappa_4 I_2^2 + \ldots . \end{aligned} \tag{3.7}$$

This expression contains two quadratic, three cubic and four quartic terms. The coefficients are called the elastic constants of the second (λ, μ), the third (ν_1, ν_2, ν_3) and the fourth $(\kappa_1, \kappa_2, \kappa_3, \kappa_4)$ order. Notations in (3.7) follow Bland (1969) but widely used are also other notations introduced by Brillouin, Murnaghan, Rivlin and others. A table presented by Truesdell and Noll (1965) gives the expressions needed for the comparison of the various third-order constants. These expressions

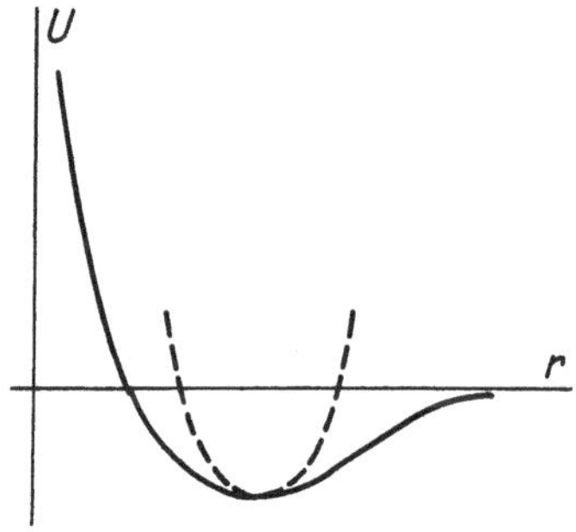

Fig. 3.1. Potential energy *vrs* distance between the atoms. Dashed line — the parabolic approximation around the minimum.

are not trivial because of the various forms of the free energy function. If only two terms in (2.7) are taken into account then the outcome is the conventional linear theory of elasticity. Other variants lead to the theory of nonlinear elasticity and the values of the coefficients are determined experimentally (Breazeale and Philip, 1984; Cantrell, 1989, etc.). The components of the stress tensor(s) can then easily be determined. For example, in 1D case

$$K_{11} = (\lambda + 2\mu)U_{1,1} + (\tfrac{1}{2}\lambda + \mu + 3\nu_1 + 3\nu_2 + 3\nu_3)U_{1,1}^2 + \dots , \tag{3.8}$$

but in 2D case, this expression turns out to be longer (with the same accuracy):

$$\begin{aligned} K_{11} = {} & (\lambda + 2\mu)U_{1,1} + (\tfrac{1}{2}\lambda + \mu + 3\nu_1 + 3\nu_2 + 3\nu_3)U_{1,1}^2 + \\ & +(6\nu_1 + 2\nu_2)U_{1,1}U_{2,2} + (\tfrac{1}{2}\lambda + 3\nu_1 + \nu_2)U_{2,2}^2 + \\ & +(\tfrac{1}{2}\lambda + \tfrac{1}{2}\nu_2 + \tfrac{3}{4}\nu_3)U_{1,2}^2 + (\tfrac{1}{2}\lambda + \mu + \tfrac{1}{2}\nu_2 + \tfrac{3}{4}\nu_3)U_{2,1}^2 + \\ & +(\nu_2 + \tfrac{3}{2}\nu_3)U_{1,2}U_{2,1} + \dots . \end{aligned} \tag{3.9}$$

The experimental determination of higher-order elastic constants has been a problem despite the growing accuracy of experiments (see Breazeale and Philip (1984) and references therein). Usually only certain combinations of constants can be determined because the changes in experimentally measurable variables are governed not by individual constants but by their combinations (like in expression (3.9)). In order to estimate the order of elastic constants, the results of Cantrell (1989) are of importance. He has shown how the elastic constants for crystals are related to the Born-Mayer hardness parameter. The calculations are based on the recurrence relations that could easily be used to find the higher-order (up to the fifth-order) elastic constants. For face-centered crystals the Brugger elastic constants of different order $C_{11}^{(2)}, C_{111}^{(3)}, C_{1111}^{(4)}, \dots$ and the Born-Mayer potential have been used (for details see Cantrell, 1989). The results have shown a thumb rule—in general (i) the magnitude of the elastic constants of each order is about a factor of ten larger

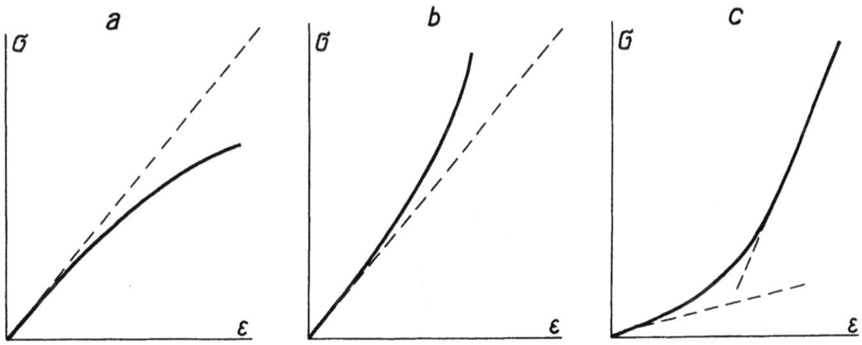

Fig. 3.2. Possible stress-strain approximations: a – linear for small deformation – soft type; b – linear for small deformations – hard type; c – nonlinear for small deformation.

than the magnitude of the previous order, and (ii) is opposite in sign. For example, in units of 10^{12} dyne cm^{-2} the elastic constants for Al crystals are:

$$C_{11}^{(2)} = 1.07\,, \quad C_{111}^{(3)} = -10.8\,, \quad C_{1111}^{(4)} = 114\,, \quad C_{11111}^{(5)} = -1280\,. \tag{3.10}$$

For polycrystalline materials, one has certainly to use this thumb rule with a certain precaution.

The series expression (3.7) is used in order to describe the deformation of compressible materials (metals, polymers, etc.) while the incompressible materials like rubber need another type of free energy (Eringen, 1962)

$$\rho_0 F = \sum_{m,n=0}^{\infty} A_{mn}(J_1 - 3)^m (J_2 - 3)^n \,, A_{00} = 0\,, \tag{3.11}$$

where A_{mn} are constants and

$$J_1 = 3 + 2I_1\,, \quad J_2 = 3 + 4I_1 + 2(I_1^2 - I_2)\,. \tag{3.12}$$

The widely used Mooney-Rivlin and Treloar potentials follow from (3.11).

It should be stressed here, that all this concerned nonlinear elasticity which is by no means an invention of contemporary science. A full account of the story of nonlinear elasticity is given by Bell (1973) showing the attempts to establish stress-strain relationships during the centuries (Mehmke, 1897). Table 3.2 illustrates the highlights of this fascinating story of ideas. Here and later, we use the notation $K_{11} = \sigma$, $U_{1,1} = \varepsilon$.

As a comment to Table 3.2 one should add a curious fact. In 1849, British Royal Iron Committee recommended that "henceforth Hooke's linear law of elasticity for iron in tension, compression, and flexure should be replaced by $\sigma = A\varepsilon - B\varepsilon^2$" (after Bell, 1973). Up to the knowledge of the present author, this recommendation has not been changed.

To sum up, all the materials are in this sense nonlinear and linearisation is usually possible only for small deformation (Fig. 3.2a,b).

Two types of materials are usually distinguished from each other: soft materials (Fig. 3.2a) and hard materials (Fig. 3.2b). Nevertheless, there are several materials

TABLE 3.2. Historical stress-strain relations (after Bell, 1973)

Functional form	Author
Series representation:	
$\varepsilon = a\sigma$	Hooke, 1678
$\varepsilon = a\sigma + \beta\sigma^2 + \gamma\sigma^3$	Thompson, 1890, metals
$\sigma = a\varepsilon + b\varepsilon^2 + c\varepsilon^3$	Cox, 1850, cast iron
$\sigma = a\varepsilon + b\varepsilon^2 + c\varepsilon^3 + d\varepsilon^4$	Hodgkinson, 1849, cast iron
Other nonlinear relations:	
$\varepsilon = a\sigma^m$	James Bernoulli, 1694
	Bülfinger, 1729
	Hodgkinson, 1822
	Bach-Schüle, 1897
$\varepsilon = \sigma(a - b\sigma)^{-1}$	Cox, 1850, cast iron
	Lang, 1896, stones, plaster
$\varepsilon = e^{m\sigma} - 1$	Imbert, 1880, India rubber
$\varepsilon = \sigma(a + be^{m\sigma})$	Poncelet, 1839, brass
$\varepsilon^2 = a\sigma^2 + b\sigma$	Wertheim, 1847, organic tissues
$\sigma = ce^{-1/\varepsilon}$	Riccati, 1731
$\sigma = c(e^{m\varepsilon} - 1)$	Hartig, 1893, leather
$\sigma = \varepsilon(1 - \varepsilon)^{-1}e^{m\varepsilon}$	Hartig, 1893, cork

behaving in a different way, for example soft tissues. Here the physical nonlinearity is of the utmost importance especially for small deformation while in the region of large deformation linear approximation is possible (Fig. 3.2c). This is caused by sliding of long molecules at the beginning of contraction and later the different structures of tissue start acting. For the corresponding potentials see Brankov (1978). However, this is already a case of a material with microstructure that must be accounted for not only in modelling soft tissues. Porosity, granular changes, phase structure, micro-cracks, etc. are the reasons for the wide scale of microstructure in media that should be taken into account. For media with regular microstructure, generalized nonlinear stress-strain relations are developed (Ostrovski, 1989), for media with irregular microstructure stochastic relations are proposed (Ostoja-Starzewski, 1991). In the latter case the microstructure is assumed to be of a granular type while the grains may be either of a soft or a hard type (c.f. Fig. 3.2a,b).

In both cases, mentioned above, however, nonlinearity may be extremely strong due to considerable changes of microstructure during the passage of wave (Nazarov et al., 1988).

One possible model for granular materials is based on the analogy to the Hertz

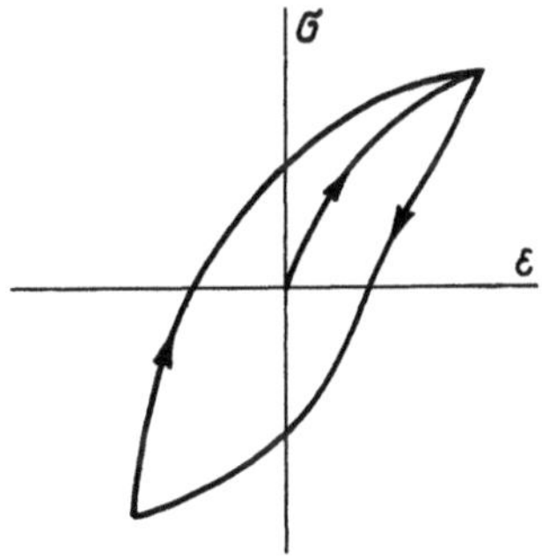

Fig. 3.3. Hysteresis.

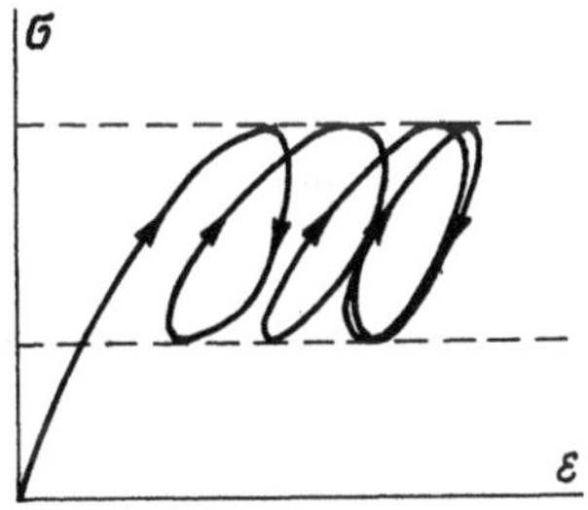

Fig. 3.4. Plastic shakedown.

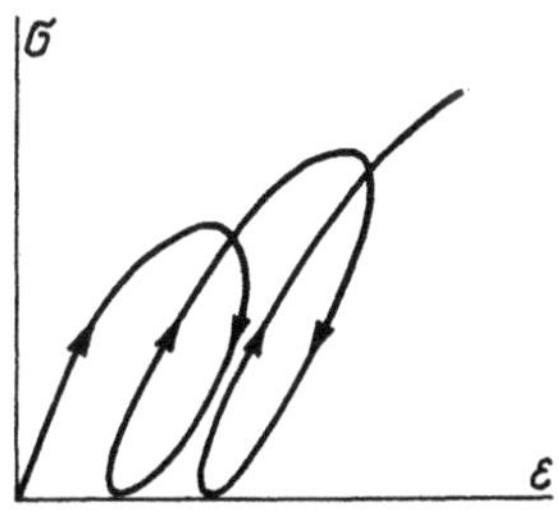

Fig. 3.5. Elastoplasticity with hardening.

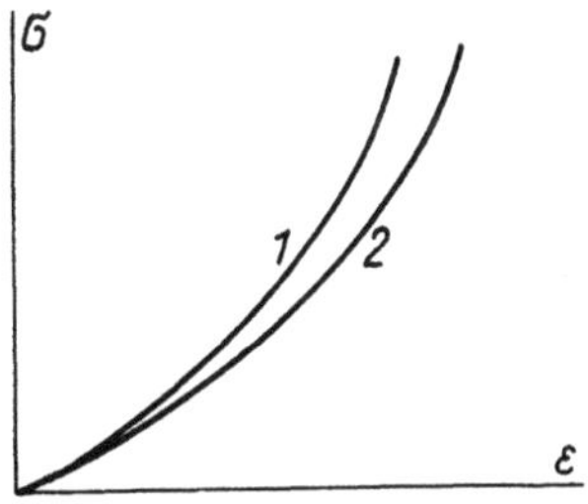

Fig. 3.6. Stress-strain curves in relaxing material: 1 – equilibrium; 2 – instantaneous.

law (Bogdanov and Skvortsov, 1992). For the longitudinal waves the Hertz law

$$F(\delta) = \gamma_1 \delta^{3/2}, \quad \gamma_1 = \text{const.} \tag{3.13}$$

gives a good description, for the transverse waves, however

$$F(\delta) = \gamma_1 \delta - \gamma_2 \mid \delta \mid \delta, \quad \gamma_1 = \text{const.}, \gamma_2 = \text{const.} \tag{3.14}$$

holds. Here F is the contact force and δ is the relative displacement.

Beyond the limit of elastic deformation, especially for ductile materials, the plastic regime is clearly exhibited. First, the hysteresis (Fig. 3.3) must be stressed with various paths of loading and unloading and then the other plastic phenomena like plastic shakedown (racheting) as shown in Fig. 3.4 and hardening effects (Fig. 3.5). For more details see Maugin (1994a).

Constitutive laws may besides the deformation-dependent variables include also the deformation rate-dependent variables (Nunziato et al., 1974; Jeffrey and Engelbrecht, 1982). It means that the long-range viscoelastic memory is taken into account and the stress is determined by the entire history of the strain. Important is here, that a nonlinear stress-strain relation in viscoelastic materials may explicitly be deformation rate-dependent and one should distinguish so-called equilibrium (static) and instantaneous (high deformation rate) curves as shown in Fig. 3.6.

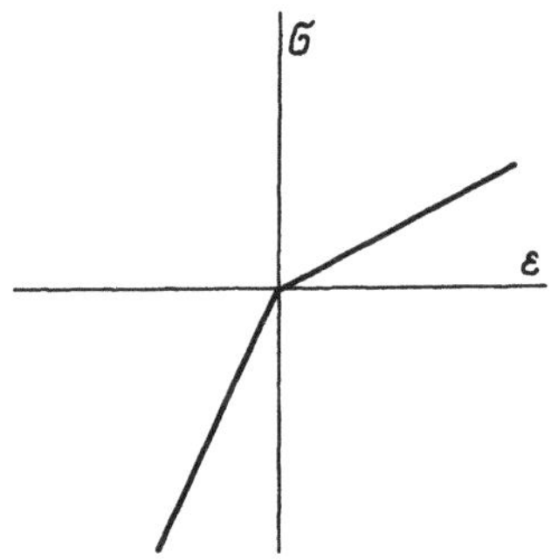

Fig. 3.7. Stress-strain relationship for bilinear materials.

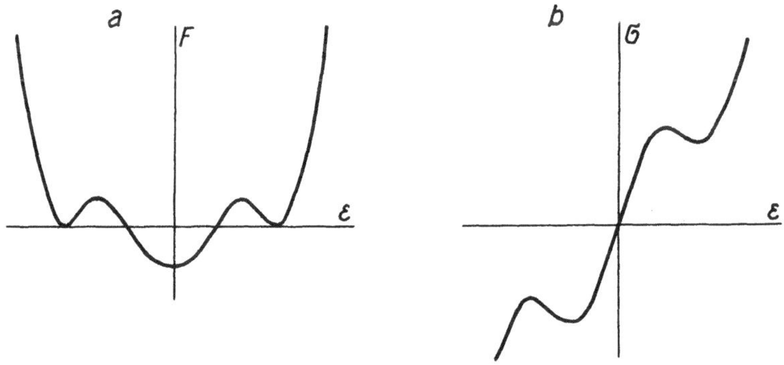

Fig. 3.8. Energy function with several minima (a) and the corresponding nonmonotonic stress-strain relationship (b).

Last but not least, two more features should be pointed out. First, materials that behave differently in compression and tension (Fig. 3.7), are also nonlinear. Waves in such media have been analysed by Maslov and Mosolov (1981). Second, the stress response functions may not always be monotonic that may happen in media with phase transitions (Pence, 1992) or in shape-memory alloys (Maugin et al., 1992). This may lead to complicated nonconvex stress-strain relations as shown in Fig. 3.8.

Nonlinearity may also be due to other constitutive laws rather than only the stress-strain relation (1.20) and one should also pay attention to relations (2.11) and (2.16). The well-known linear Fourier law can have the form

$$Q_i = k(T)T_{,i}\,, \tag{3.15}$$

where $k(T)$ is the temperature-dependent function. In addition, the supply of the energy in (2.6) may be temperature-dependent. The outcome—the equation of heat conduction is then (Kudryumov et al., 1988)

$$\dot{T} = (k(T)T_{,i})_{,i} + q_0 T^{\beta} \tag{3.16}$$

where $q_0 = \text{const.}$, $\beta = \text{const.}$

Finally, the body force can be related to deformation like

$$f^k = \sum_{n=0} A_n^k U_{i,i}^n \tag{3.17}$$

where $A_n^k = \text{const.}$ Such an expression is used for modelling long waves in media with microstructure (Engelbrecht and Khamidullin, 1988).

Here we have listed nonlinearities for observable variables. Internal variables and their nonlinearities is a new and promising area of research (Maugin, 1990).

3.3. OTHER TYPES OF NONLINEARITIES

3.3.1. Geometrical nonlinearities

Intuition says that this type of nonlinearity must be related to large geometrical changes. First of all, this is reflected in the exact expressions of the strain tensors. For the Green deformation tensor E_{ij} (or E_{KL}), the basic expression is

$$E_{ij} = \frac{1}{2}(U_{i,j} + U_{j,i} + U_{k,j}U_{i,k}). \tag{3.18}$$

This expression is always used in the theory of large deformations contrary to the theory of infinitesimal deformations (linear theory). In terms of space (Eulerian) coordinates the Cauchy deformation tensor e_{ij} reads

$$e_{ij} = \frac{1}{2}(u_{i,j} + u_{j,i} - u_{k,j}u_{i.k}) \tag{3.19}$$

There is no controversy in signs of (3.18) and (3.19) because of the different strain measures. Physically it means that the free energy function in terms of E_{ij} or e_{ij} is also different resulting in same physical effects.

Finite albeit small deformations should be taken into account not only in rubber-type materials but also in metal-type materials besides the other nonlinearities. There is an important point to be stressed. For usual metal-type materials the geometrical nonlinearity of the type (3.18) must always be combined with the physical nonlinearities. The reason is the following. The physical nonlinearity for those materials is usually of the soft type and much stronger than the geometrical nonlinearity. Introducing only the geometrical nonlinearity, the result will be physically inadmissible. For example, a geometrically nonlinear theory in this sense permits then the tensile shock waves contrary to the compression shock waves according to the theory involving both geometrical and physical nonlinearities.

Turning to Eq. (2.4), there are more geometrical nonlinearities involved that are related to the terms like $x^k_{,L}$. Since space and material coordinates differ by the displacement then it should be considered as an additional source of nonlinearities, especially in problems with curved coordinates (Eringen, 1962). It is easy to see that as

$$x_i = X_i + U_i\,, \tag{3.20}$$

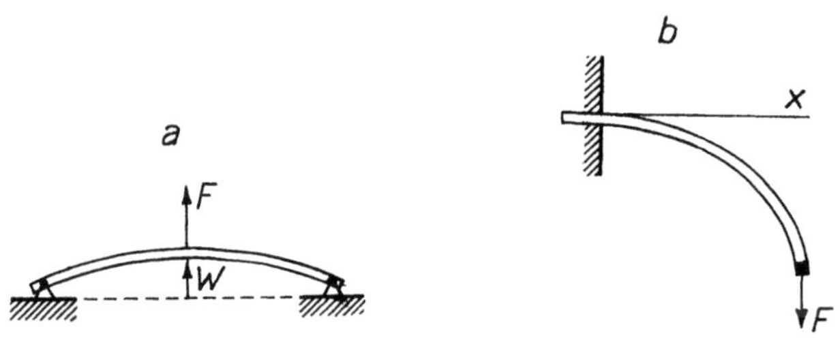

Fig. 3.9. Large deflections: a – a beam; b – a console.

then

$$\frac{\partial x_i}{\partial X_j} = \delta_{ij} + U_{i,j} \tag{3.21}$$

where δ_{ij} is the Kronecker symbol.

If the mathematical models are adjusted to structural elements like plates, beams, etc., then the geometrical character of large displacements comes into account replacing continuum mechanics approach by approximated theories. For example, in Fig. 3.9 two typical cases are shown. For the beam depicted in Fig. 3.9a, the force F is a nonlinear function in terms of the deflection w

$$F = k_1 w + k_3 w^3\,, \quad k_1 = \text{const.}\,, \quad k_3 = \text{const.} \tag{3.22}$$

and for the console in Fig. 3.9b, the bending moment M is expressed as

$$M = \frac{ku''}{[1+(u')^2]^{3/2}}\,, \tag{3.23}$$

where $u(x)$ is the displacement and $(\ )'$ denotes the differentiation with respect to the argument. The geometrical nonlinearity is crucial for loop solitons in flexible strings as shown in Fig. 1.3.

3.3.2. Kinematical nonlinearities

The motion is a cause for additional nonlinear effects. In fluids, the convective terms in the Navier-Stokes' equations $(v_x \cdot \partial v_x/\partial x, \ldots)$ are well understood. This nonlinearity arises from the dual role of the velocity in determining the acceleration of a fluid particle dealing D/Dt as the substantive derivative (Tritton, 1988).

In the compound motion of solids, local coordinate systems are used to describe motion relative to some inertial reference frame. The absolute acceleration $\mathbf{A}$ is given then

$$\mathbf{A} = \mathbf{a} + \mathbf{A_0} + \dot{\mathbf{\Omega}} \times \mathbf{r} + \mathbf{\Omega} \times \mathbf{\Omega} \times \mathbf{r} + 2\mathbf{\Omega} \times \mathbf{v}\,, \tag{3.24}$$

where $\mathbf{a}$ is the acceleration in the local frame, $\mathbf{A_0}$ that of the origin of the local frame, $\mathbf{\Omega}$ is the angular velocity relative to the reference frame, $\mathbf{r}$ and $\mathbf{v}$ are the local position vector and velocity, respectively. Last three terms in (3.24) are nonlinear in variables $\mathbf{v}$ and $\mathbf{r}$.

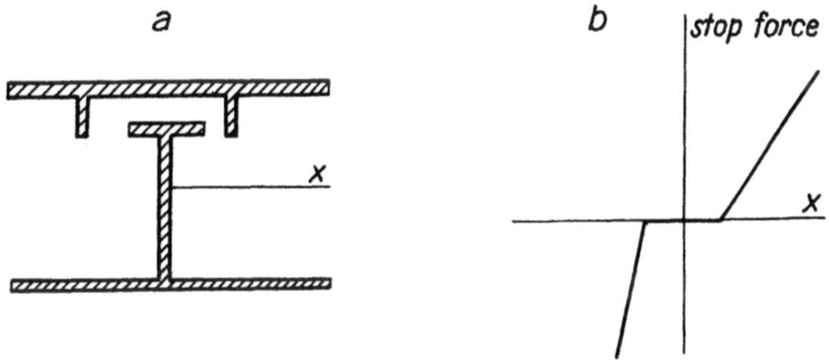

Fig. 3.10. Motion of a beam with an elastic constraint: a – physical scheme, b – force-deformation relation (after Crandall, 1974).

In rotation dynamics, the nonlinear terms appear in the Euler's equations for a rigid body in pure rotation. These equations are of the type

$$M_x = I_x\dot{\omega}_x - (I_z - I_y)\omega_y\omega_z \, , \tag{3.25}$$

where M_x is the applied moment with respect to x-axis, I_x, I_y, I_z are the principal second moments of mass and $\omega_x, \omega_y, \omega_z$ are the components of the angular velocity. It should be pointed out that this is important for wave propagation in the twisted elastica (Davies and Moon, 1993) due to the analogy between the free spinning rigid body and the elastica.

3.3.3. Structural nonlinearities

In the engineering practice, nonlinearities may also be caused by specific properties of the elements composing the whole structure. According to Crandall (1974), these are called structural nonlinearities. Two examples are shown in Fig. 3.10 and 3.11 (Crandall, 1974). The first shows how the force-deformation relation can be piece-wise linear, i.e. globally nonlinear; the second — how the force-deformation relation is nonlinear for a system with springs. In terms of continuum mechanics, these examples can be tackled as problems with constraints.

A special structural nonlinearity arises also due to Coulomb friction (Fig. 3.12) where the effective damping parameter is discontinuous between stick and slip.

3.3.4. Combined nonlinearities

The terminology is not established and these nonlinearities could also be named coupling nonlinearities. In some sense, this set of nonlinearities is closely related to the previous sets but some new aspects are stressed here.

First, physical and geometrical nonlinearities are always coupled, either through stress-strain relationships or through $T^{KL}x^{l}_{,K}$ – type expressions in (2.4). One should notice here that for usual constructive materials the geometrical nonlinearity in E_{ij} can not be taken into account without physical nonlinearities (see explanations in Engelbrecht, 1983).

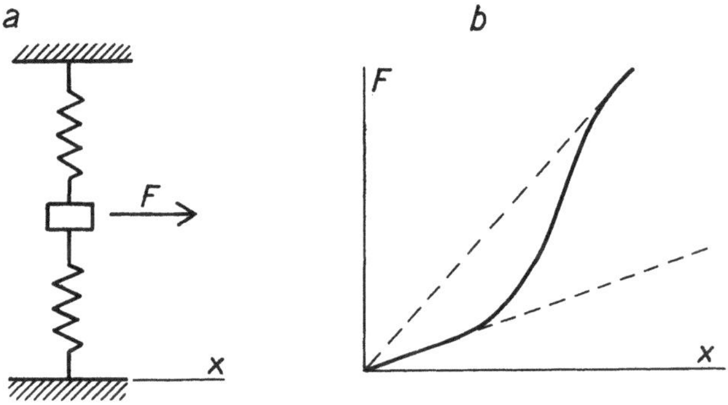

Fig. 3.11. System with springs: a – physical scheme, b – force-deformation relation (after Crandall, 1974).

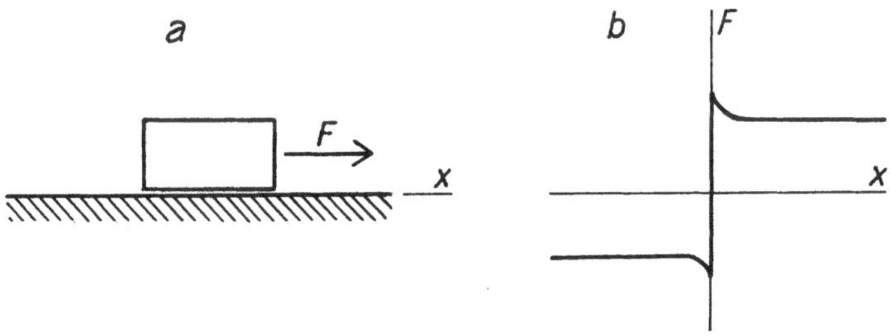

Fig. 3.12. Coulomb friction: a – physical scheme; b – friction force *vrs* slip velocity (after Crandall, 1974).

Next, nonlinear coupling between predeformation and generated wave fields is to be taken into account in modern acoustodiagnostics (Guz, 1986). The same goes for the coupling between inhomogeneities and generated fields (Ravasoo, 1989) that affects both the phase velocities and the amplitudes of all the harmonics.

In general terms, the temperature effects combined with mechanical waves may also be nonlinear. Usually the temperature changes are considered around the equilibrium temperature T_0 so that

$$|T - T_0|/T_0 \ll 1 \tag{3.26}$$

is satisfied. If not then the conservation of energy contains also combined nonlinearities. Equation (1.17) yields

$$\rho_0 T \left(\frac{\partial^2 F}{\partial T \partial E_{KL}} \dot{E}_{KL} + \frac{\partial^2 F}{\partial T^2} \dot{T} \right) +_D T^{KL} \dot{E}_{KL} + Q^K_{,K} + \rho_0 h = 0\,, \tag{3.27}$$

where ${}_D T^{KL}$ denotes the viscous part of the stress tensor. Assuming here the derivatives with respect to T constant, there are still nonlinearities like $T\dot{E}_{KL}$ and

$T\dot{T}$, that should be taken into account. For example, we might get (Engelbrecht, 1983)

$$T\kappa\dot{U}_{1,1} + TT_0^{-1}c_E\dot{T} = Q_{1,1} \tag{3.28}$$

with $\kappa =$ const., $c_E =$ const.

Nonlinear Rayleigh waves are characterized by a special type of nonlinearity that could be named nonlocal (Hamilton et al., 1995a). There are two reasons for that:

(i) nonlinearity involves interaction between horizontal and vertical components of a wave; these components are related through Hilbert transform;

(ii) interaction caused by the frequency-dependent penetration depth of a Rayleigh wave. The corresponding evolution equation will be presented later (Section 5.1).

One interesting example of combined nonlinearities is the nonlinear coupling of longitudinal and transverse waves in solids. Bland (1969) has shown that the transverse deformation cannot exist without the longitudinal deformation because of the different speeds. It is especially obvious in the two-dimensional case (Peipman et al., 1992).

3.4. REMARKS

One could certainly ask now: "What is the use of nonlinearities?" Should one avoid them as cumbersome complications or vice versa, should one try to benefit from them? At this stage we proceed with the analysis and return to this question later, in Chapter 7.

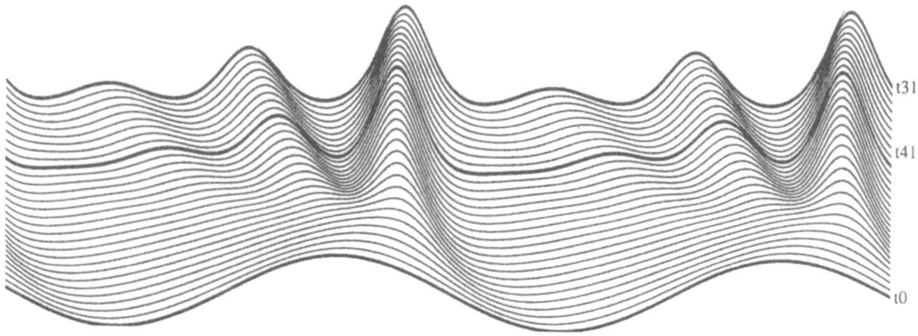

4. Nonlinear wave dynamics: mathematical models

One of the principal objects of theoretical research in any department of knowledge is to find the point of view from which the subject appears in its greatest simplicity.
J.W.Gibbs

4.1. GOVERNING EQUATIONS

4.1.1. Mathematical models

We consider the system of equations and relations like (2.17) or (2.19) to be a concrete mathematical model if the Helmholtz free energy function F and all the other dependent variables (the irreversible part of the stress tensor ${}_DT^{KL}$, for instance) are determined by a finite number of terms and their coefficients. For that, i.e. for the justification of series representation and truncation of series with a necessary accuracy auxiliary assumptions are needed.

For a wide class of structural materials the auxiliary postulates given above are justified only when the following auxiliary assumptions are satisfied (Eringen, 1962)

$$U_{I,J} \ll 1\,, \tag{4.1}$$

$$| T - T_0 | \, T_0^{-1} \ll 1\,. \tag{4.2}$$

These assumptions permit us to represent the Helmholtz free energy F in terms of, say, Taylor series. Assumption (4.2), however, is strict and leads to essential limitations with regard to the change of temperature. Another form of the auxiliary assumption for temperature proposes that temperature change is bounded by

$$\int_{T_0}^{T} \alpha_T dT \sim | U_{I,J} |\,, \tag{4.3}$$

where α_T is the coefficient of linear thermal expansion. This assumption requires that we should also take into account the mechanical and thermal inhomogeneities that arise as a result of the greater temperature change. Since here we focus our attention on deformation waves, assumption (4.2) is considered to be the fundamental one. The Helmholtz free energy function must be approximated in physical units to obtain also the components of stress tensors in physical components. The Taylor series representation then gives

$$\begin{aligned} F(E_{KL},T) &= F(0,T_0)+\frac{\partial F(0,T_0)}{\partial E_{KL}}E_{KL}+\frac{\partial F(0,T_0)}{\partial T}T+ \\ &+\frac{1}{2}\frac{\partial^2 F(0,T_0)}{\partial E_{KL}\partial E_{IJ}}E_{KL}E_{IJ}+\frac{\partial^2 F(0,T_0)}{\partial E_{KL}\partial T}E_{KL}(T-T_0)+ \\ &+\frac{1}{2}\frac{\partial^2 F(0,T_0)}{\partial T^2}(T-T_0)^2+\cdots\,, \end{aligned} \tag{4.4}$$

where E_{KL} denotes the physical components of the deformation tensor.

The Helmholtz free energy F is now taken in a definite form with the fixed number of terms in the series (4.4). Let us assume that

$$\rho_0 F=\tfrac{1}{2}\lambda I_1^2+\mu I_2+\nu_1 I_1^3+\nu_2 I_1 I_2+\nu_3 I_3-\kappa_{1t}I_1(T-T_0)-\kappa_{2t}(T-T_0)^2 \tag{4.5}$$

i.e. the usual form of the third-order expansion (Bland, 1969) which is called the Murnaghan potential. Here λ,μ are the Lamé coefficients and ν_i, $i=1,2,3$, the third-order elastic constants, and

$$\kappa_{1t}=(3\lambda+2\mu)\alpha_T\,,\quad \kappa_{2t}=\frac{1}{2}\rho_0 c_E T_0^{-1} \tag{4.6}$$

where c_E is the specific heat per unit mass. There are several ways of constructing (4.5) and the third-order elastic constants are denoted by various letters (c.f. Section 3.2). Here we follow the general approach given by Bland (1969), changing only the notation.

The dissipative part of the stress tensor T^{KL} is given by (2.14).

Now final mathematical models can be easily derived and several examples are given below.

3D system, rectangular coordinates:

$$\begin{aligned} &\rho_0\ddot{U}_N-C_{NKLM}(U_{I,J})U_{K,LM}-D_{NKLM}(U_{I,J})\ddot{U}_{K,LM}= \\ &\quad =B_{NK}T_{,K}+A_{LMN}T_{,LM}\,, \\ &g\dot{T}+H_{MN}\dot{U}_{M,N}=Q_{K,K}\,, \\ &\tau_0\dot{Q}_N+Q_N=kT_{,N}+R_{NKL}\dot{U}_{K,L}\,. \end{aligned} \tag{4.7}$$

The coefficients of this system are given by Engelbrecht (1983).

2D system, rectangular coordinates:

$$\begin{aligned} &[\lambda+2\mu+3(\lambda+2\mu+2\nu_1+2\nu_2+2\nu_3)U_{1,1}+\lambda+6\nu_1+2\nu_2)U_{2,2}]U_{1,11}+ \\ &\quad+[(2\lambda+4\mu+2\nu_2+3\nu_3)U_{1,2}+(2\mu+2\nu_2+3\nu_3)U_{2,1}]U_{1,12}+ \\ &\quad+[\mu+(\lambda+2\mu+\nu_2+\tfrac{3}{2}\nu_3)U_{1,1}+(\lambda+2\mu+\nu_2+\tfrac{3}{2}\nu_3)U_{2,2}]U_{1,22}+ \\ &\quad+[(\lambda+2\mu+\nu_2+\tfrac{3}{2}\nu_3)U_{1,2}+(\mu+\nu_2+\tfrac{3}{2}\nu_3)U_{2,1}]U_{2,22}+ \\ &\quad+[\lambda+\mu+(\lambda+\mu+6\nu_1+3\nu_2+\tfrac{3}{2}\nu_3)U_{1,1}+ \\ &\quad+(\lambda+\mu+6\nu_1+3\nu_2+\tfrac{3}{2}\nu_3)U_{2,2}]U_{2,21}+ \\ &\quad+[(\mu+\nu_2+\frac{3}{2}\nu_3)U_{1,2}+(\lambda+2\mu+\nu_2+\tfrac{3}{2}\nu_3)U_{2,1}]U_{2,11}+ \end{aligned}$$

$$+ (\xi_\nu + \tfrac{4}{3}\eta_\nu)\dot{U}_{1,11} + \eta_\nu(\dot{U}_{1,22} + \dot{U}_{2,12}) - \kappa_{1t}(T - T_0)(U_{1,11} + U_{1,22}) -$$
$$- \kappa_{1t}(1 + U_{1,1})T_{,1} - \kappa_{1t}U_{1,2}T_{,2} - \rho_0\ddot{U}_1 = 0\,,$$
$$[(\lambda + 2\mu + \nu_2 + \tfrac{3}{2}\nu_3)U_{2,1} + (\mu + \nu_2 + \tfrac{3}{2}\nu_3)U_{1,2}]U_{1,11} +$$
$$+ [\lambda + \mu + (\lambda + \mu + 6\nu_1 + 3\nu_2 + \tfrac{3}{2}\nu_3)U_{1,1} +$$
$$+ (\lambda + \mu + 6\nu_1 + 3\nu_2 + \tfrac{3}{2}\nu_3)U_{2,2}]U_{1,12} +$$
$$+ [(\lambda + 2\mu + 2\nu_2 + \tfrac{3}{2}\nu_3)U_{1,2} + (\mu + \nu_2 + \tfrac{3}{2}\nu_3)U_{2,1}]U_{1,22} +$$
$$+ [\lambda + 2\mu + 3(\lambda + 2\mu + 2\nu_1 + 2\nu_2 + 2\nu_3)U_{2,2} + (\lambda + 6\nu_1 + 2\nu_2)U_{1,1}]U_{2,22} +$$
$$+ [(2\mu + 2\nu_2 + 3\nu_3)U_{1,2} + (2\lambda + 4\mu + 2\nu_2 + 3\nu_3)U_{2,1}]U_{2,21} +$$
$$+ [\mu + (\lambda + 2\mu + \nu_2 + \tfrac{3}{2}\nu_3)U_{1,1} + (\lambda + 2\mu + \nu_2 + \tfrac{3}{2}\nu_3)U_{2,2}]U_{2,11} +$$
$$+ \eta_\nu(\dot{U}_{1,12} + \dot{U}_{2,11}) + (\xi_\nu + \tfrac{4}{3}\eta_\nu)\dot{U}_{2,22} - \kappa_{1t}(T - T_0)(U_{2,22} + U_{2,11}) -$$
$$- \kappa_{1t}U_{2,1}T_{,1} - \kappa_{1t}(1 + U_{1,1})T_{,2} - \rho_0\ddot{U}_2 = 0\,,$$
$$T\kappa_{1t}\dot{U}_{1,1} + T\kappa_{1t}\dot{U}_{2,2} + TT_0^{-1}\rho_0 c_E\dot{T} = Q_{1,1} + Q_{2,2}\,,$$
$$\tau_0\dot{Q}_1 + Q_1 = kT_{,1}\,,$$
$$\tau_0\dot{Q}_2 + Q_2 = kT_{,2}\,. \tag{4.8}$$

1D model, rectangular coordinates:

$$(\lambda + 2\mu)[1 + 3(1 + m_0)U_{1,1} - \kappa(\lambda + 2\mu)^{-1}(T - T_0)]U_{1,11} +$$
$$+ (\xi_\nu + \tfrac{4}{3}\eta_\nu)\dot{U}_{1,11} - \kappa_{1t}(1 + U_{1,1})T_{,1} - \rho_0\ddot{U}_1 = 0\,,$$
$$T\kappa_{1t}\dot{U}_{1,1} + TT_0^{-1}\rho_0 c_E\dot{T} = Q_{1,1}\,,$$
$$\tau_0\dot{Q}_1 + Q_1 = kT_{,1}\,. \tag{4.9}$$

Here

$$m_0 = 2(\nu_1 + \nu_2 + \nu_3)(\lambda + 2\mu)^{-1} \tag{4.10}$$

determines the measure of the physical nonlinearity. The coefficient $1 + m_0$ determines the coupled influence of geometrical and physical nonlinearities. The two cases $m_0 > 0$ and $m_0 < 0$ are possible. As $| m_0 | = 1 \ldots 10$ for the usual constructive materials, the simultaneous consideration of geometric and physical nonlinearities is justified.

1D model, spherical coordinates:

$$[\lambda + 2\mu + 3(\lambda + 2\mu + 2\nu_1 + 2\nu_2 + 2\nu_3)U_{1,1} +$$
$$+ 2(\lambda + 6\nu_1 + 2\nu_2)X_1^{-1}U_1 - \kappa(T - T_0)]U_{1,11} +$$
$$+ 2[\lambda + 2\mu + (2\lambda + 2\mu + 6\nu_1 + 4\nu_2 + 3\nu_3)U_{1,1} +$$
$$+ (6\nu_1 + 2\nu_2)X_1^{-1}U_1]X_1^{-1}U_{1,1} -$$
$$- 2[\lambda + 2\mu + (3\lambda + 3\mu + 18\nu_1 + 3\nu_2)X_1^{-1}U_1]X_1^{-2}U_1 +$$
$$+ (\xi_\nu + \tfrac{4}{3}\eta_\nu)(\dot{U}_{1,11} + 2X_1^{-1}\dot{U}_{1,1} - 2X_1^{-2}\dot{U}_1) - \kappa_{1t}(1 + U_{1,1})T_{,1} -$$
$$- \kappa_{1t}2X_1^{-1}(T - T_0)U_{1,1} - \rho_0\ddot{U}_1 = 0\,,$$
$$T\kappa_{1t}\dot{U}_{1,1} + 2T\kappa X_1^{-1}\dot{U}_1 + TT_0^{-1}\rho_0 c_E\dot{T} = Q_{1,1} + 2X_1^{-1}Q_1\,,$$

$$\tau_0 \dot{Q}_1 + Q_1 = kT_{,1}\,. \tag{4.11}$$

All these mathematical models are derived from (2.17). As an example, let us also consider relaxing material (Eringen, 1962; Jeffrey and Engelbrecht, 1982) when the kernel G^{KLIJ} in (2.18) is given by an exponential function. Then we have for the 1D case and rectangular coordinates the following governing equation:

$$(\lambda + 2\mu)[1 + 3(1 + m_0)U_{1,1}]U_{1,11} + \\ + \varepsilon_1(\lambda + 2\mu)\left[\int_0^t \frac{\partial U_{1,1}}{\partial \tau} \exp\left(-\frac{t-\tau}{\tau_0}\right) d\tau\right]_{,1} - \rho_0 \ddot{U}_1 = 0\,, \tag{4.12}$$

where ε_1, τ_0 are constants.

When predeformation is to be taken into account, the outcome turns out to be very complicated. Let us propose that the initial state of the viscoelastic medium is an undeformed natural state. At the instant $t = 0$ the medium is subjected to the external forces and further on it is deformed. At the instant $t = t_0 > 0$ the wave process in the predeformed medium is excited so that

$$U_K^*(X_J, t) = U_K^0(X_J, t) + U_K(X_J t) \tag{4.13}$$

where U_K^* denotes the displacement at the present instant, U_K^0 is caused by predeformation and U_K corresponds to the wave motion. Even in a simplified case of plane inhomogeneous predeformation ($U_3^* = U_3^0 = 0$) the outcome is quite cumbersome. The quasi-onedimensional longitudinal wave is, for example, described by the equation of motion (Engelbrecht and Ravasoo, 1994)

$$U_{1,11} + f_1(X_1, t)U_{1,1} - f_2(X_1, t)[R(t) * U_{1,11}] - f_3(X_1, t)\ddot{U}_1 = \\ = f_4(X_1, t)U_1 U_{1,11}\,, \tag{4.14}$$

where the conventional notation of the convolution integral is used

$$F_1(X, t) * F_2(X, t) = \int_0^t F_1(X, t-\tau)F_2(X, \tau)d\tau\,. \tag{4.15}$$

The coordinate X_2 may be regarded here as a parameter and $R(t)$ is the kernel function. The functions f_i in eq. (4.14) are given in terms of the initial displacement:

$$f_0(X_1, t) = 1 + k_1 U_{1,1}^0(X_1, t) + k_2 U_{2,2}^0(X_1, t)\,, \\ f_1(X_1, t) = f_0^{-1}(X_1, t)[k_1 U_{1,11}^0(X_1, t) + k_3 U_{1,22}^0(X_1, t) + k_4 U_{2,21}^0(X_1, t)]\,, \\ f_2(X_1, t) = k_0 f_0^{-1}(X_1, t) > 0\,, \\ f_3(X_1, t) = \rho_0 f_2(X_1, t) > 0\,, \\ f_4(X_1, t) = -k_1 f_0^{-1}(X_1, t)\,. \tag{4.16}$$

The coefficients k_i are related to the second and the third order elastic constants, ρ_0 is the density. Such mathematical models have been used for solving the problems in the nondestructive testing (Ravasoo, 1995).

4.1.2. Mathematical models in matrix form

The matrix formulation is widely used in reductive methods. Here we represent the mathematical models derived in Section 4.1.1 in the matrix notation which allows a more compact description. For the general model (4.7) the following vector of dependent variables is introduced

$$\mathbf{U} = \begin{vmatrix} \dot{U}_N \\ U_{K,L} \\ T \\ Q_K \end{vmatrix}, \quad N, K, L = 1, 2, 3. \tag{4.17}$$

System (4.7) yields then

$$I\frac{\partial \mathbf{U}}{\partial t} + A^K \frac{\partial \mathbf{U}}{\partial X_K} + \sum_{p=2}^{q} B_M^{rs} \frac{\partial^p \mathbf{U}}{\partial (X_M)^r \partial t^s} + \mathbf{H} = 0, \tag{4.18}$$

where $r + s = p$ and

$$A^K = A^K(\mathbf{U}), \quad B_M^{rs} = B_M^{rs}(\mathbf{U}), \quad \mathbf{H} = \mathbf{H}(\mathbf{U}). \tag{4.19}$$

Later we see that this concise form allows most effectively to separate various nonlinearities. In 1D case, Eq. (4.18) is

$$I\frac{\partial \mathbf{U}}{\partial t} + A\frac{\partial \mathbf{U}}{\partial X} + \sum_{p=2}^{q} B^{rs} \frac{\partial^p \mathbf{U}}{\partial X^r \partial t^s} + \mathbf{H} = 0, \tag{4.20}$$

where $A^1 = A, X_1 = X, B_1^{rs} = B^{rs}$.

It is possible that the vector of dependent variables is presented in a different way. For example, Peipman et al. (1992) have used for the 2D case the following vectors (theory of viscoelasticity):

$$\mathbf{V} = \begin{vmatrix} \dot{U}_1 \\ U_{1,1} \\ U_{1,2} \end{vmatrix}, \quad \mathbf{W} = \begin{vmatrix} \dot{U}_2 \\ U_{2,1} \\ U_{2,2} \end{vmatrix}, \tag{4.21}$$

and Eq. (4.18) yields

$$\begin{aligned} &I\frac{\partial \mathbf{V}}{\partial t} + P^1\frac{\partial \mathbf{V}}{\partial X_1} + P^2\frac{\partial \mathbf{V}}{\partial X_2} + P^3\frac{\partial \mathbf{W}}{\partial X_1} + P^4\frac{\partial \mathbf{W}}{\partial X_2} + \\ &\quad + Q^1\frac{\partial^2 \mathbf{V}}{\partial t \partial X_1} + Q^2\frac{\partial^2 \mathbf{V}}{\partial t \partial X_2} + Q^3\frac{\partial^2 \mathbf{W}}{\partial t \partial X_1} + Q^4\frac{\partial^2 \mathbf{W}}{\partial t \partial X_2} = 0, \\ &I\frac{\partial \mathbf{W}}{\partial t} + R^1\frac{\partial \mathbf{V}}{\partial X_1} + R^2\frac{\partial \mathbf{V}}{\partial X_2} + R^3\frac{\partial \mathbf{W}}{\partial X_1} + R^4\frac{\partial \mathbf{W}}{\partial X_2} + \\ &\quad + S^1\frac{\partial^2 \mathbf{V}}{\partial t \partial X_1} + S^2\frac{\partial^2 \mathbf{V}}{\partial t \partial X_2} + S^3\frac{\partial^2 \mathbf{W}}{\partial t \partial X_1} + S^4\frac{\partial^2 \mathbf{W}}{\partial t \partial X_2} = 0, \end{aligned} \tag{4.22}$$

where P^i, Q^i, R^i, S^i, $i = 1, 2, 3, 4$ are the matrices dependent on $\mathbf{V}$ and $\mathbf{W}$. This form was used in order to analyse the nonlinear coupling of longitudinal and transverse waves.

4.2. EVOLUTION EQUATIONS

4.2.1. Introduction

The governing equations for wave motion are, as a rule, derived from the conservation laws of continuous media. In simpler cases these equations are strictly hyperbolic but usually the physical effects which should be taken into account are so complicated that various asymptotics bring us to less complicated systems (Whitham, 1974). The effects of nonlinearity certainly form one source of complexity but the effects of dispersion, damping, inhomogeneity, etc. should also be taken into consideration. As a result, the governing equations (see above) are not integrable by analytic methods. This is the reason why a lot of attention is paid to seeking the methods of simplification which should lead to the less complex equations but retain all the important physical effects.

The problem is not only in the mathematical complexity of governing equations but also in the number of waves. If in the linear theory, for example, longitudinal and shear waves can be easily separated, then in the nonlinear theory the coupling can affect both waves considerably. In a general case of a complicated system of equations the main question is to understand to which wave which physical effects are related both quantitatively and qualitatively.

One of the possibilities to overcome such difficulties in contemporary mathematical physics is to introduce the notion of evolution equations governing just one single wave. Physically it means the separation (if possible) of a multi-wave process into separate waves. The waves are then governed by the so-called evolution equations every one of which describes the distortion of a single wave along a properly chosen characteristics (ray). Most of the well-known equations describing solitary waves belong to this class of equations. The main idea of constructing such "simplified" (with regard to the initial multi-wave system) equations is the following. A set of small parameters related either to the initial conditions or to the physical and/or geometrical parameters which form the coefficients of the initial equations, is introduced and the perturbation method together with the method of stretched coordinates is then applied. Taniuti and his co-workers who have initiated such an approach (Taniuti and Nishihara, 1983) called it "reductive perturbation method". Actually there are several methods which are used to simplify the governing equations (Engelbrecht et al., 1988). These are:

(i) asymptotic (reductive perturbation) method;
(ii) iterative method;
(iii) spectral method;
(iv) method of approximate (etalon) equations.

Here the main attention will be paid to the method (i), others will be discussed briefly.

4.2.2. The asymptotic (reductive perturbation) method

Basic assumptions. We use the matrix notation like in Section 4.1.2 and the system under consideration is written in the following form

$$I\frac{\partial \mathbf{U}}{\partial t} + A^K \frac{\partial \mathbf{U}}{\partial X^K} + \sum \varepsilon^{m(p)} B_{rs}^{\alpha\beta} \frac{\partial^p \mathbf{U}}{\partial (X^\alpha)^r \partial (X^\beta)^s} + \mathbf{H} = 0\,, \tag{4.23}$$

where $\mathbf{U}$ is an n-vector, $K = 1,2,3$; $\alpha,\beta = 0,1,2,3$; $X^0 = t$, $r + s = p \geq 2$, $m \geq 1$. The coefficients may depend on X^K and $\mathbf{U}(X^K, t)$, being functions of the geometrical and physical parameters of the medium. Vector $\mathbf{H}$ may also contain the integral operators of $\mathbf{U}$ and in many cases it contains the small parameter ε. Equation (4.23) is quasilinear because $A^K = A^K(\mathbf{U})$ and with a parabolic part (higher derivatives due to the dissipative and/or dispersive character of the wave process). However, the wave-type character of the process leads to small parameters $\varepsilon^{m(p)}$ before the higher derivatives with $m(p) \geq 1$.

The construction of evolution equations means actually an asymptotic analysis of Eq. (4.23) with initial and boundary conditions

$$\mathbf{U}(X^K, t) \mid_{t=0} = \mathbf{\Psi}(X^K); \quad \mathbf{U}(X^K, t) \mid_S = \mathbf{\Phi}(X^\alpha), \tag{4.24}$$

where S is a determined contour.

Assumption 1. *The terms in equation (4.23) are smooth functions of* $\mathbf{U}$.

Assumption 2. *It is possible to develop vector* $\mathbf{U}$ *into the power series in a small parameter*

$$\mathbf{U} = \mathbf{U}_0 + \varepsilon \mathbf{U}_1 + \varepsilon^2 \mathbf{U}_2 + \dots . \tag{4.25}$$

Assumption 2 is not restrictive in principle and it is needed for the correctness of the sequential procedure. It is possible to construct the evolution equation of the first order without this assumption but then it is difficult to give the physical meaning to the terms of the higher order in ε. This is the common approach in nonlinear acoustics (Naugolnykh and Ostrovsky, 1990). Next we introduce

Assumption 3. *It is possible to develop matrices* $A^K, B^{\alpha\beta}_{rs}$ *and vector* $\mathbf{H}$ *into power series in a small parameter:*

$$\begin{aligned} A^K &= A_0^K(X^M) + \varepsilon A_1^K(X^M, \mathbf{U}) + \dots , \\ B^{\alpha\beta}_{rs} &= B^{\alpha\beta}_{0rs}(X^K) + \varepsilon B^{\alpha\beta}_{1rs}(X^K, \mathbf{U}) + \dots , \\ \mathbf{H} &= \mathbf{H}_0 + \varepsilon \mathbf{H}_1 + \dots . \end{aligned} \tag{4.26}$$

Substituting (4.25) into Eq. (4.23) we get

$$I \frac{\partial \mathbf{U}_0}{\partial t} + A^K \frac{\partial \mathbf{U}_0}{\partial X^K} + \mathbf{H} = 0 \tag{4.27}$$

and, using also (4.4)

$$I \frac{\partial \mathbf{U}_0}{\partial t} + A_0^K \frac{\partial \mathbf{U}_0}{\partial X^K} + \mathbf{H}_0 = 0 . \tag{4.28}$$

Definition 1: *Equation (system) (4.23) is basic.*

Definition 2: *Equation (system) (4.27) is quasilinear associated.*

Definition 3: *Equation (system (4.28) is linear associated.*

Now the final assumption follows:

Assumption 4. *The eigenvalues of* A_0^K *are real, distinct and non-degenerate.*

For the linear associated equation (4.28) the usual ray method may be used. The wave front is determined by the eikonal equation

$$t = \varphi(X^K) \tag{4.29}$$

that actually determines the points at time t on the wave front. The eikonal φ is usually determined by the equation

$$\mid \operatorname{grad} \varphi \mid^2 = \lambda_i^{-2} \tag{4.30}$$

where λ_i is the velocity of the wave under consideration.

Basic principle. We start with the following definitions:

Definition 4. *The wave vector* $\mathbf{U}(X^\alpha)$ *describes a simple wave if there exists a family of propagating surfaces,* $\xi = t - \varphi(X^K)$, *such that the magnitude of the rate of change of* $\mathbf{U}(X^\alpha)$ *or of its derivatives is zero if* X^K *is moving with this surface.*

D e f i n i t i o n 5. *The wave vector* $\mathbf{U}(X^\alpha)$ *describes a progressive wave if there exists a family of propagating surfaces,* $\xi = t - \varphi(X^K)$, *such that the magnitude of the rate of change of* $\mathbf{U}(X^\alpha)$ *or of its derivatives is small as compared with the magnitude of the rate of change of* $\mathbf{U}(X^\alpha)$ *when* X^K *is fixed.*

In simple wave theory it is possible to use

$$\mathbf{U}(X^K, t) = \sum_n \phi_n(X^K) f_n(\xi), \quad \xi = t - \varphi(X^K), \tag{4.31}$$

where the phase functions f_n satisfy certain conditions of differentiation. The short-wave asymptotics (ray method) includes

$$f_n = (i\omega)^{-n} \exp(i\omega\xi), \quad n = 0, 1, 2, \ldots \tag{4.32}$$

In this case the traditional Helmholtz equation leads in the first approximation to the eikonal equation (4.30) and in the next approximations to the evolution equations

$$2\nabla\phi_n \nabla\varphi + \phi_n \nabla^2\varphi = 0, \quad n = 0, 1, 2, \ldots \tag{4.33}$$

The evolution equations (4.33) are ordinary differential equations where the derivatives are taken with respect to a parameter along the ray.

In the theory of progressive waves the basic equation (4.23) does not permit a series representation of the form (4.31) because of distortions which arise in the wave profile with increasing time (or with the increase of the coordinate that plays the role of time). Taking into account Assumptions 1–4 and Definition 5, the basic principle for finding asymptotic solutions of the basic equation may be formulated.

B a s i c p r i n c i p l e. *The evolution equations of the basic system in terms of the amplitude factor along the correspondent rays are constructed on the basis of wave front kinematics of the associated linear system.*

From Definition 5 we get:

C o r o l l a r y 1. *Distortions of a progressive wave with respect to space coordinates are of order* $O(1)$ *and, with respect to phase coordinate* ξ, *of order* $O(\varepsilon)$.

Kinematics. The ray coordinates are presented in agreement with the method of stretched coordinates (Cole, 1968):

(i) without taking into account interaction,

$$\xi = \varepsilon^k(t - \varphi(X^K)), \quad \tau^\nu = \varepsilon^{k+1+m(\nu)} X^\nu; \tag{4.34}$$

(ii) taking into account interaction,

$$\xi = \varepsilon^k[t - \varphi(X^K) + \varepsilon\phi_j(X^\alpha)], \quad \tau^\nu = \varepsilon^{k+1+m(\nu)} X^\nu, \tag{4.35}$$

where k, m are real numbers and ϕ_j is a phase function.

Let us check Corollary 1. For simplicity, let $m = 0$. If $\partial/\partial t = O(\varepsilon)$, $\partial/\partial X^K = O(\varepsilon)$, then $\partial/\partial\xi = O(\varepsilon^{1-k})$, $\partial/\partial\tau^\nu = O(\varepsilon^{-k})$. Here we have assumed $\partial\varphi/\partial X^K = O(1)$, $\partial\xi/\partial X^\alpha = O(1)$. If $k = 0$, then $\partial/\partial\xi = O(\varepsilon)$, $\partial/\partial\tau^\nu = O(1)$.

The index k gives the space scale with respect to the wave front. The first point is to balance the nonlinear and the dissipative and/or dispersive effects. To keep these terms of the same order, the relation

$$k + 1 = pk + m(p) \tag{4.36}$$

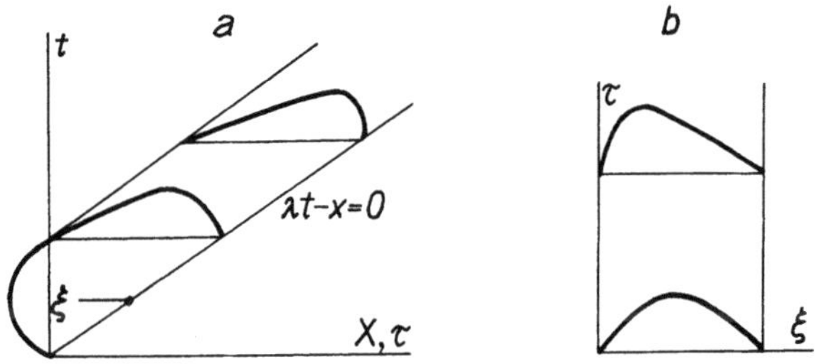

Fig. 4.1. Homogeneous medium: a – X, t plane; b – ξ, τ plane.

must be satisfied.

In the one-dimensional case for the homogeneous medium the ray coordinates may be simplified. Let $X^1 \equiv X$ then we have either

$$\xi = \varepsilon^k(\lambda_i t - X) \quad \tau = \varepsilon^{k+1} X \tag{4.37}$$

or

$$\xi = \varepsilon^k(\lambda_i t - X) \,, \quad \tau = \varepsilon^{k+1} t \,, \tag{4.38}$$

The τ coordinates in (4.37) and (4.38) have different dimensions and therefore we introduce

D e f i n i t i o n 6: *Transformation (4.37) is of the space-space kind.*

D e f i n i t i o n 7: *Transformation (4.38) is of the space-time kind.*

For the inhomogeneous medium the situation is more complicated because from the eikonal equation it follows that time t is dependent on $\lambda_k(x)$. Hence, the ray coordinates are

$$\xi = \varepsilon^k \left(t - \int_0^X \frac{dx}{\lambda_k(x)} \right) \,, \quad \tau = \varepsilon^{k+1} X \,. \tag{4.39}$$

D e f i n i t i o n 8. *Transformation (4.39) is of the time-space kind.*

In the multidimensional case there are more possibilities of stretching the coordinates (4.34). Two cases may be distinguished here:

(i) τ^ν contains only space coordinates ($\nu = 1, 2, 3$);

(ii) τ^ν contains both time and space coordinates ($\nu_1 = 0 \,, \ \nu_{2,3} \in 1, 2, 3$).

The wave front is curved in X^K space and the determination of the eikonal is a complicated problem. However, several simplifications are possible, from which the most important is the case of a plane front approximation. Here usually the form

$$\xi = \varepsilon^k(\lambda_i t - X^i) \,, \quad \tau^\nu = \varepsilon^{k+1+m(\nu)} X^\nu \tag{4.40}$$

is introduced where the parameter $m(\nu)$ is matched with the change rates of the components of vector $\mathbf{U}$ (see below).

Two examples of the physical planes $X^1 = X, t$ and $\xi, \tau^1 = \tau$ are shown in Fig. 4.1 and Fig. 4.2.

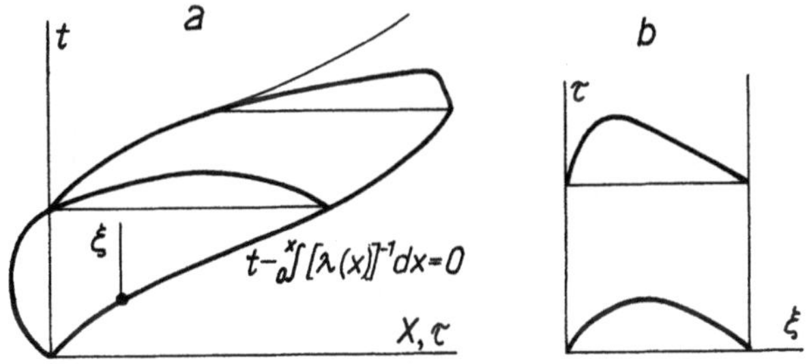

Fig. 4.2. Inhomogeneous medium: a – X,t plane; b – ξ,τ plane.

Evolution equations. We start from one-dimensional cases.

C a u c h y p r o b l e m, h o m o g e n e o u s m e d i u m. From (4.2) the following conditions follow

$$\boldsymbol{\Psi}(X) \neq 0\,, \quad \boldsymbol{\Phi}(t) = 0\,. \tag{4.41}$$

Here and further $X^1 \equiv X$. The associated system (4.28) yields a sequence of eigenvalues, i.e. velocities λ_i, $i = 1, 2, \dots, n$. We follow one progressive wave labelled λ_j. The space-time transformation

$$\xi = \lambda_j t - X\,, \quad \tau = \varepsilon t \tag{4.42}$$

is used. Substituting (4.25), (4.26) and (4.42) into Eq. (4.23) we get the following sequence of operator equations of various powers in ε:

$$L_0^t\left(\frac{\partial \mathbf{U}_0}{\partial \xi}\right) + \mathbf{H}_0 = 0\,, \tag{4.43a}$$

$$L_1^t\left(\frac{\partial \mathbf{U}_1}{\partial \xi}\right) + M_0^t(\mathbf{U}_0) + \mathbf{H}_1 = 0\,, \tag{4.43b}$$

$$L_2^t\left(\frac{\partial \mathbf{U}_2}{\partial \xi}\right) + M_1^t(\mathbf{U}_1) + N_0^t(\mathbf{U}_0) + \mathbf{H}_2 = 0\,, \tag{4.43c}$$

...

where, for $n = 0, 1, 2, \dots$,

$$L_n^t\left(\frac{\partial \mathbf{U}_n}{\partial \xi}\right) = (I\lambda_j - A_0)\frac{\partial \mathbf{U}_n}{\partial \xi}\,,$$

$$M_n^t(\mathbf{U}_n) = I\lambda_j\frac{\partial \mathbf{U}_n}{\partial \tau} - A_1\frac{\partial \mathbf{U}_n}{\partial \xi} - \sum_p B_{0rs}(\lambda_j)^s\frac{\partial^p \mathbf{U}_n}{\partial \xi^p}\,, \tag{4.44}$$

...

We determine the left and the right eigenvectors $\mathbf{l}_j$ and $\mathbf{r}_j$, respectively, from the equation

$$\mathbf{l}_j(I\lambda_j - A_0) = (I\lambda_j - A_0)\mathbf{r}_j = 0 \tag{4.45}$$

with the normalizing condition $\mathbf{l}_j \cdot \mathbf{r}_j = 1$.

The solution of Eq. (4.43a) is

$$\mathbf{U}_0 = \alpha_0(\xi, \tau)\mathbf{m}\,. \tag{4.46}$$

D e f i n i t i o n 9. *In (4.46),* $\alpha_0(\xi, \tau)$ *is an amplitude factor.*

D e f i n i t i o n 10. *In (4.46),* $\mathbf{m}$ *is a structure factor.*

Equation (4.43a), containing the differentiation with respect to ξ, permits determination of the structure factor but not the amplitude factor. If $\mathbf{H}_0 = 0$, then $\mathbf{m} = \mathbf{r}_j$, i.e. the first approximation of the unknown vector $\mathbf{U}_0$ is decomposed into the components of $\mathbf{r}_j$. We use this simplified variant to achieve more concise results. Now we multiply Eq. (4.43b) by the normalized left eigenvector $\mathbf{l}_j$ and get

$$\frac{\partial \alpha_0}{\partial \tau} + a_1 \alpha_0 \frac{\partial \alpha_0}{\partial \xi} + \sum_p a_{2p} \frac{\partial^p \alpha_0}{\partial \xi^p} + f_0(\alpha_0) = 0\,, \tag{4.47}$$

$$a_1 = \lambda_j^{-1} \mathbf{l}_j A_1(\mathbf{r}_j)\mathbf{r}_j\,, \quad f_0(a_0) = \lambda_j^{-1} \mathbf{l}_j \mathbf{H}_1(\mathbf{r}_j)\alpha_0\,,$$

$$a_{2p} = (-1)^n (\lambda_j)^{s-1} \mathbf{l}_j B_{0rs} \mathbf{r}_j\,. \tag{4.48}$$

D e f i n i t i o n 11. *Equation (4.47) is an evolution equation of the first order.*

This describes the evolution of the wave propagating along the characteristics $\lambda_j X - t = \text{const}$. It is a partial differential equation (c.f. (4.33)) to be solved under the following initial condition

$$\alpha_0(\xi, \tau)\,|_{t=\tau_0} = \mathbf{l}_j \cdot \mathbf{\Psi}(\xi)\,. \tag{4.49}$$

An evolution equation of the second order can easily be derived from (4.43c).

B o u n d a r y p r o b l e m, h o m o g e n e o u s m e d i u m. Here we have $\mathbf{\Psi}(X) = 0$, $\mathbf{\Phi}(t) \neq 0$. The transformation here is of the space-space kind:

$$\xi = \lambda_j t - X\,, \quad \tau = \varepsilon X\,. \tag{4.50}$$

The sequence of equations of various powers in ε is

$$L_0^X \left(\frac{\partial \mathbf{U}_0}{\partial \xi} \right) + \mathbf{H}_0 = 0\,, \tag{4.51a}$$

$$L_1^X \left(\frac{\partial \mathbf{U}_1}{\partial \xi} \right) + M_0^X(\mathbf{U}_0) + \mathbf{H}_1 = 0\,, \tag{4.51b}$$

$$L_2^X \left(\frac{\partial \mathbf{U}_2}{\partial \xi} \right) + M_1^X(\mathbf{U}_1) + N_0^X(\mathbf{U}_0) + \mathbf{H}_2 = 0\,, \tag{4.51c}$$

...

where

$$L_n^X \left(\frac{\partial \mathbf{U}_n}{\partial \xi} \right) = L_n^t \left(\frac{\partial \mathbf{U}_n}{\partial \xi} \right) = (I\lambda_j - A_0) \frac{\partial \mathbf{U}_n}{\partial \xi}\,,$$

$$M_n^X(\mathbf{U}_n) = A_0 \frac{\partial \mathbf{U}_n}{\partial \tau} - A_1 \frac{\partial \mathbf{U}_n}{\partial \xi} + \sum_p B_{0rs}(-1)^r (\lambda_j)^s \frac{\partial^p \mathbf{U}_n}{\partial \xi^p}\,, \tag{4.52}$$

...

Solution (4.46) holds and, taking into account decomposition into the components of the right eigenvector $\mathbf{r}_j$, the evolution equation of the first order gives, from (4.51b),

$$\frac{\partial \alpha_0}{\partial \tau} + a_1 \alpha_0 \frac{\partial \alpha_0}{\partial \xi} + \sum_p a_{2p} \frac{\partial^p \alpha_0}{\partial \xi^p} + f_0(\alpha_0) = 0\,, \tag{4.53}$$

$$a_1 = \lambda_j^{-1} \mathbf{l}_j A_1(\mathbf{r}_j)\mathbf{r}_j, \quad a_{2p} = (-1)^r (\lambda_j)^{s-1} \mathbf{l}_j B_{0rs} \mathbf{r}_j\,,$$

$$f_0(\alpha_0) = \lambda_j^{-1} \mathbf{l}_j \mathbf{H}_1(\mathbf{r}_j)\alpha_0\,, \tag{4.54}$$

with the initial condition

$$\alpha_0(\xi,\tau)\mid_{\tau=\tau_0} = -\mathbf{l}_j \cdot \mathbf{\Phi}(\xi)\,. \tag{4.55}$$

Comparison of (4.49) and (4.55) leads to:

C o r o l l a r y 2. *The Cauchy problem for an evolution equation is obtained using a space-time transformation for the Cauchy problem of the basic system and a space-space transformation for the boundary value problem of the basic system.*

Our next target is a multidimensional system. Several cases are analysed by Engelbrecht (1983), here we note only a most instructive case where the various stretching is used. The vector of field variables is presented as

$$\mathbf{U} = \mathbf{V} + \mathbf{W} \tag{4.56}$$

(c.f. (3.34)) and then the series

$$\mathbf{V} = \mathbf{V}_0 + \varepsilon\mathbf{V}_1 + \dots\,,$$
$$\mathbf{W} = \varepsilon^t(\mathbf{W}_0 + \varepsilon\mathbf{W}_1 + \dots) \tag{4.57}$$

are introduced. Using the transformation

$$\xi = \lambda_j t - X^j\,, \quad \tau^\nu = \varepsilon^{1+m} X^\nu\,, \tag{4.58}$$

the final outcome depends upon the choice of t and m. An instructive case, used widely in nonlinear acoustics gives a two-dimensional evolution equation in the form

$$\frac{\partial}{\partial\xi}\left[\frac{\partial\alpha_0}{\partial t} + \alpha_1\alpha_0\frac{\partial\alpha_0}{\partial\xi} + \sum a_{2p}\frac{\partial^p\alpha_0}{\partial\xi^p} + f_0(\alpha_0)\right] = \Delta\alpha_0\,, \tag{4.59}$$

where Δ is the Laplacian in the transverse direction. Actually this equation describes the distortion of a bounded wave beam when the diffractional expansion is taken into account. The derivation of such equations can be found in Naugolnykh and Ostrovsky (1990).

The general theory of the asymptotic (reductive perturbation) method is illuminated by Taniuti and Nishihara (1983), and also by Engelbrecht (1983).

To sum up, the following operations are needed in order to use the asymptotic (reductive perturbation) method:

- eigenvalues and eigenvectors of A_0 found;
- stretched coordinates (moving frame) introduced:
- the series interpretation introduced;
- perturbation method applied;
- evolution equations derived.

4.2.3. Other methods

The iterative method. We present the system (4.23) in the following form (one-dimensional case):

$$I\frac{\partial\mathbf{U}}{\partial t} + A(\tau,\chi)\frac{\partial\mathbf{U}}{\partial X} = \varepsilon\mathbf{F}(\mathbf{U},X,t,\tau,\chi,\varepsilon)\,, \tag{4.60}$$

where $\tau = \varepsilon t, \chi = \varepsilon X$ and all the nonlinearities are in the right-hand side. In case $\varepsilon = 0$ the system (4.60) has the solution

$$\mathbf{U} = \sum_{i=1}^{n} v_i(X - \lambda_i t)\mathbf{r}_i\,, \tag{4.61}$$

where v_i is a scalar function and $\mathbf{r}_i$, as above, the right eigenvector of matrix A.

We represent the initial system (4.60) in the normal form (Courant and Hilbert, 1963) changing the variable $\mathbf{U}$ to $\mathbf{V}$ by making use of the formula

$$\mathbf{U} = Y\mathbf{V}, \tag{4.62}$$

where Y is the $n \times n$ matrix formed by the linearly independent eigenvectors $\mathbf{r}_i$ of matrix A. As far as A is dependent on "slow" coordinates τ and χ, the eigenvectors $\mathbf{r}_i$ and the eigenvalues λ_i depend also on these coordinates. Substituting (4.62) into (4.60) we get

$$\frac{\partial \mathbf{V}}{\partial t} + \theta\frac{\partial \mathbf{V}}{\partial X} = \varepsilon Y^{-1}\left[\mathbf{F} - \left(\frac{\partial Y}{\partial \tau} + A\frac{\partial Y}{\partial \chi}\right)\mathbf{V}\right] = \varepsilon f[\mathbf{V}, X, t, \chi, \tau, \varepsilon]. \tag{4.63}$$

Here θ is the diagonal matrix with the elements $\lambda(\tau,\chi)$ and Y^{-1} is the inverse matrix of Y, consisting of the left eigenvectors $\mathbf{l}_i$ of A. The usual normalization is here assumed: $\mathbf{l}_i\mathbf{r}_j = \delta_{ij}$; δ_{ij} is the Kronecker symbol.

According to (4.62) and (4.63) the solution may be given as a superposition of waves, each one propagating (at $\varepsilon = 0$) along its characteristics and interacting with others only due to the perturbation $f = 0$. In many cases only $m < n$ waves v_i are generated by the initial and boundary conditions. When the initial condition is localized then the wave process is soon decomposed into the separate independent waves, so that in far field each of them may be analysed separately. In this case vector $\mathbf{V}$ contains m "basic"waves and other $n - m$ waves may be calculated by the usual perturbation scheme.

The main subcase is $m = 1$. The asymptotic procedure (see Engelbrecht et al., 1988) gives

$$\frac{\partial v}{\partial t} + \lambda(\tau,\chi)\frac{\partial v}{\partial X} = \varepsilon f(v, v^{(n-1)}, \chi, \tau, \varepsilon), \tag{4.64}$$

which is the sought evolution equation. The method is proposed by Gorschkov et al. (1974).

The summary is the following:

- system (4.60) solved with $\varepsilon = 0$ and eigenvalues and eigenvectors found;
- system (4.60) transformed to the normal form;
- decomposition into "basic" and "other" waves;
- evolution equations derived.

The spectral method. This method is convenient for waves in dispersive media and gives an excellent possibility to explain the decomposition into single waves (Miropolski, 1981). The system (4.23) is again rewritten in a generalized form

$$I\frac{\partial \mathbf{U}}{\partial t} + \mathbf{LU} = \varepsilon\mathbf{MU}, \tag{4.65}$$

where $\mathbf{L}$ is a linear differential operator (a matrix depending on $\partial\mathbf{U}/\partial X$), $\mathbf{M}$ is a nonlinear operator. The wave process is decomposed into the eigenvalues of the linear ($\varepsilon = 0$) problem

$$\mathbf{U} = \sum_m \int dk U_k^m \mathbf{r}_k^m \exp(i(\omega_k^m t - kX)), \tag{4.66}$$

where $\mathbf{r}_k^m$ is determined by

$$\mathbf{L}(-ik)\mathbf{r}_k^m = -i\omega_k^m \mathbf{r}_k^m\,,$$
$$\det|\,\mathbf{L}(-ik) + i\omega_k^m I\,| = 0\,. \tag{4.67}$$

Here U_k^m is the spectral amplitude, and m is the mode number. We assume $U_k^m = U_k^m(t)$. Substituting (4.66) into (4.65) and performing the inverse Fourier transform in the coordinate, we get the infinite system of equations for the spectral amplitudes

$$\frac{dU_k^m}{dt} = \varepsilon \sum_{m_1,m_2} \int dk_1 dk_2 V_{k\,k_1\,k_2}^{mm_1m_2} U_{k_1}^{m_1} U_{k_2}^{m_1} \times$$
$$\times \exp[i(\omega_{k_1}^{m_1} + \omega_{k_2}^{m_2} - \omega_k^m)t\;\delta(k_1 + k_2 - k)]\,, \tag{4.68}$$

where the matrix-coefficients $V_{k\,k_1\,k_2}^{mm_1m_2}$ are determined by the given operator $\mathbf{M}$ and depend only upon the wave number k and the mode numbers m.

The decomposition into single waves is possible if

$$V_{k\,k_1\,k_2}^{mm_1m_2} = 0\,, \quad m_{1,2} \neq m \tag{4.69}$$

and in this case the wave with wave number m is propagating without any dependence on other modes. The equation (4.68) yields then (index m is neglected)

$$\frac{dU_k}{dt} = \int dk_1 dk_2 V_{kk_1k_2} U_{k_1} U_{k_2} \exp(i\Delta t)\delta(k_1 + k_2 - k)\,. \tag{4.70}$$

Now the Fourier transform must be performed in order to obtain the evolution equation that, generally speaking, may be of an integro-differential form. The details of this method can be found in Engelbrecht et al. (1988).

The procedures needed here are consequently the following:

- decomposition with respect to eigenvalues;
- derivation of equations for spectral amplitudes;
- decomposition checked;
- evolution equation restored by inverse transform.

The method of approximate (etalon) evolution equations. The basic idea is given by Whitham (1974) for a linear problem. Let us consider the one-dimensional evolution equation

$$\frac{\partial u}{\partial \tau} + \int_{-\infty}^{\infty} K(\xi - z)\frac{\partial u}{\partial z}dz = 0\,, \tag{4.71}$$

where τ and ξ are the independent variables, u is a certain wave variable (particle velocity, deformation, etc.) and the kernel function $K(z)$ describes the dispersion properties. This equation has elementary solutions of the form

$$u = u_0 \exp(ik_*\xi + i\omega_*\tau)\,, \quad u_0 = \text{const.}\,, \tag{4.72}$$

provided the dispersion relation

$$c_{ph} = \omega_* k_*^{-1} = \int_{-\infty}^{\infty} K(\xi)\exp(ik_*\xi)d\xi \tag{4.73}$$

is satisfied. As the right hand side of expression (4.73) is the Fourier transform of the given kernel $K(\xi)$, the inverse transform

$$K(\xi) = \frac{1}{2\pi}\int_{-\infty}^{\infty} c(k_*)\exp(ik_*\xi)dk_* \tag{4.74}$$

allows us to determine the kernel for every given phase velocity $c_{ph}(k)$. The trivial example of deriving the linearized Korteweg-de Vries (KdV) equation according to this scheme is well-known. In this case the phase velocity has a polynomial form with even terms in it, and the kernel function is given by the delta-function and its consecutive derivatives (Whitham, 1974). It is appropriate that some remarks based on physical ideas should now be made. First, the independent variables in an evolution equation correspond to a moving frame, i.e., usually $\xi = c_0 t - x$, $\tau = \varepsilon x$ where ε is a small parameter and $c_0 = \text{const.}$ is the "translation" velocity. Usually c_0 is either the phase velocity c_{ph} determined from the corresponding linear associated problem or the sound velocity close to c_{ph}. Secondly, the wave number k_* and the frequency ω_* in (4.72) correspond to the variables ξ and τ. In real space-time coordinates x and t the wave number k and the frequency ω are different from those. It is easily concluded that

$$k_* = -\omega c_0^{-1}\,, \quad \omega_* = (k - \omega c_0^{-1})\varepsilon^{-1}\,. \tag{4.75}$$

Thirdly, evolution equations are usually derived for weakly dispersive systems. In the case of strongly dispersive system, the group velocities must be used instead of the phase velocities when constructing the moving frame.

Based on these comments, and the expressions given above, a simple asymptotic approach for deriving evolution equations may be developed.

Suppose we know the dispersion relation

$$G(\omega, k) = 0 \tag{4.76}$$

for a certain wave. Let an asymptotic expression

$$G_e(\omega, k) = 0 \tag{4.77}$$

exist provided the additional condition

$$f(\omega, k) = 0 \tag{4.78}$$

is satisfied. Condition(s) (4.78) may be easily interpreted physically; for example, it might characterize high or low frequency processes. The main idea is to determine the kernel function in (4.74) not from the exact dispersion relation (4.76) but from its asymptotic expression (4.77). This may considerably simplify the evaluation of the inverse transform (4.74) and, consequently, may result in an explicit form of the kernel function which often cannot be obtained by using the exact dispersion relation. We denote the kernel function corresponding to the asymptotic expression (4.77) by $K_e(\xi)$. The evolution equation

$$\frac{\partial v}{\partial \tau} + \int_{-\infty}^{\infty} K_e(\xi - z)\frac{\partial v}{\partial z} dz = 0 \tag{4.79}$$

describes the process provided the condition (4.78) is satisfied, and will be called the *etalon evolution equation.* Notice once more that the difference between k_*, ω_* and k, ω must be clearly stated in order to obtain physically admissible results.

The evolution equations are usually derived from the system of equations of motion. Matching the physical and mathematical requirements, it is easily concluded that the 1st-order evolution equation describes the situation where the dispersive, nonlinear, dissipative e.a. terms on the level of the evolution equation are additive (Whitham, 1974). The solutions to these nonlinear evolution equations are by no means additive, reflecting effectively the possible balance between different physical effects. All these ideas may be summed up in the following successive scheme for constructing nonlinear evolution equations:

- derive or measure the general dispersion relation (4.76);
- derive the asymptotic dispersive relation (4.55) under the condition (4.78);
- determine the kernel $K_e(\xi)$ from (4.74);
- construct the linear evolution equation (4.79);
- add the nonlinear term according to the nonlinear wave motion in the corresponding nondispersive medium;
- get the final nonlinear etalon evolution equation which describes the process under conditions (4.78).

4.3. MODELLING OF WAVES IN INERTIAL-DIFFUSIVE SYSTEMS

In Section 2.4 the notion of internal variables was introduced and analysed. In this case a mathematical model involving both observable and internal variables is of the mixed type. The observable variables are governed by equations of hyperbolic type at least in the sense described in Section 4.2.2. The internal variables in their turn are governed by the kinetic equations, i.e. by the parabolic equations. Below we present a rather general treatment of such a mixed problem.

Let an n-vector $\mathbf{U}$ be the vector of observable variables (c.f. Section 2.4) and $\mathbf{W}$—an m-vector of internal variables. For the sake of simplicity we deal only with a 1D system $X_1 = X, t$. The dynamics of such a process is then described by a system of $n + m$ equations

$$I\frac{\partial \mathbf{U}}{\partial t} + A\frac{\partial \mathbf{U}}{\partial X} + \varepsilon B_{11}\frac{\partial^2 \mathbf{U}}{\partial X \partial t} + \ldots = \mathbf{H}(\mathbf{U}, \mathbf{W})\,, \tag{4.80}$$

$$I\frac{\partial \mathbf{W}}{\partial t} + D_{11}\frac{\partial^2 \mathbf{W}}{\partial X^2} + \ldots = \mathbf{P}(\mathbf{U}, \mathbf{W})\,, \tag{4.81}$$

where as before I is a unit matrix; A, B_{11}, D_{11} are certain matrices of order $O(1)$ and $\mathbf{H}, \mathbf{P}$ are the coupling vectors. If $m = 1$, then we have

$$I\frac{\partial \mathbf{U}}{\partial t} + A\frac{\partial \mathbf{U}}{\partial X} + \varepsilon B_{11}\frac{\partial^2 \mathbf{U}}{\partial X \partial t} + \ldots = \mathbf{H}(\mathbf{U}, w)\,, \tag{4.82}$$

$$\frac{\partial w}{\partial t} + d_{11}\frac{\partial^2 w}{\partial X^2} + \ldots = p(u_j, w)\,. \tag{4.83}$$

The initial conditions for both systems (4.80), (4.81) and (4.82), (4.83) should be consistent

$$\mathbf{U}(X,t)\,|_{t=0} = \phi(X), \quad \mathbf{U}(X,t)|_{X=0} = \psi(t)\,, \tag{4.84}$$

$$w(X,t)\,|_{t=0} = w(X)\,, \tag{4.85}$$

and, if needed, some boundary conditions specified for w (Grindrod,1991).

If the system is uncoupled, i.e. if $\mathbf{H} = \mathbf{H}(\mathbf{U}), \mathbf{P} = \mathbf{P}(\mathbf{W})$ or $p = p(w)$, then the dissipative (reaction-diffusion) subsystems (4.81) or (4.83) may still lead to wave-type solutions provided their r.h.s. have certain properties (Kolmogorov et al., 1937; Fischer, 1937).

The main problem here is to find how dynamics of observable variables modelled by a hyperbolic equation (or by an equation with a hyperbolic main part) is influenced by internal variables governed by diffusion-type (parabolic) equations. This seems to be an important problem in many areas of physics. Maugin (1990) has shown how to build up the mathematical models for nematic crystals, localization of damage and plastic strain. The thermodynamical viewpoint of nerve pulse transmission was briefly analysed by Maugin and Engelbrecht (1994). However, up to the knowledge of the author there is no special treatment proposed concerning the analysis of complex systems like (4.80), (4.81) or (4.82), (4.83) which reflect different properties of the processes although the problem is acute (Engelbrecht, 1989). A special case of nerve pulse transmission using a system like (4.82) and (4.83) will be analysed later in Chapter 6.

Here we propose a simple straightforward approach for deriving an evolution equation governing the behaviour of observable variables. Let us take (4.82), (4.83) as basic. Then the following operations should be carried out:

- construct the evolution equation on the basis of (4.82) for a certain $u_i \in \mathbf{U}$ (c.f. Section 4.2) treating w as a parameter;
- transform the scalar equation (4.83) to moving frame coordinates τ, ξ (c.f. Section 4.2);
- analyse this simplified system.

If, for example, from (4.82) we get in terms of $\alpha \sim u_i$

$$\frac{\partial \alpha}{\partial t} + a_1 \alpha \frac{\partial \alpha}{\partial \xi} + a_2 \frac{\partial^2 \alpha}{\partial \xi^2} + \ldots = h(\alpha, w) \tag{4.86}$$

and from the scalar equation (4.83) just

$$\frac{\partial w}{\partial \xi} = p(\alpha, w)\,, \tag{4.87}$$

then, if $h(\alpha, w)$ is of the 1st order in w, we get

$$\frac{\partial}{\partial \xi}\left(\frac{\partial \alpha}{\partial t} + a_1 \alpha \frac{\partial \alpha}{\partial \xi} + a_2 \frac{\partial^2 \alpha}{\partial \xi^2} + \ldots\right) = \frac{\partial h}{\partial w}\frac{\partial w}{\partial \xi} + \frac{\partial h}{\partial \alpha}\frac{\partial \alpha}{\partial \xi}\,. \tag{4.88}$$

A similar equation was obtained for nerve pulse transmission (Engelbrecht, 1981) where the leading term was also a mixed derivative $\partial^2 \alpha / \partial \xi \partial \tau$. In comparison with the traditional evolution equation, this means a more complicated situation accounting for one internal variable.

4.4. NONLINEARITIES REVISITED

In the analysis three levels of nonlinearities must be separated corresponding to the different stages of analysis:

(i) *basic level*: the level of basic equations (conservation laws, constitutive laws, deformation, ...; see Chapter 2);

(ii) *derived level*: the level of the mathematical model — quasilinearity of

$$A^K \frac{\partial \mathbf{U}}{\partial X^K}\,, \mathbf{H}\,, \ldots\,. \tag{4.89}$$

(iii) *approximation level*: the level of evolution equations involving gradient nonlinearity

$$v^p \frac{\partial v}{\partial \xi}\,, \quad p \geq 1\,, \tag{4.90}$$

direct nonlinearity

$$v^q\,, \quad q \geq 2 \tag{4.91}$$

and coupling nonlinearity (see Chapter 4). Here $p = 1$ corresponds to the quadratic nonlinearity, $p = 2$ to the cubic nonlinearity, etc.

It seems to be useful to give a simple example based on a 1D elastic wave motion (longitudinal waves). Nonlinearities of the *basic* level are explicitly seen from the stress-strain relation

$$K_{11} = (\lambda + 2\mu)U_{1,1} + \left(\tfrac{1}{2}\lambda + \mu + 3\nu_1 + 3\nu_2 + 3\nu_3\right) U_{1,1}^2\,. \tag{4.92}$$

The second term in this expression emphasizes the influence of nonlinearity. The physical nonlinearity is expressed by the term

$$(3\gamma_1 + 3\gamma_2 + 3\gamma_3)U_{1,1}^2 \tag{4.93}$$

and the geometrical nonlinearity—by the term

$$\left(\tfrac{1}{2}\lambda + \mu\right) U_{1,1}^2\,. \tag{4.94}$$

The latter is due to the nonlinear term in the deformation tensor

$$E_{11} = U_{1,1} + \tfrac{1}{2}U_{1,1}^2\,. \tag{4.95}$$

The *derived* level is related to the equation

$$I\frac{\partial \mathbf{U}}{\partial t} + A^1\frac{\partial \mathbf{U}}{\partial X^1} = 0\,, \tag{4.96}$$

$$\mathbf{U} = \begin{vmatrix} \dot{U}_1 \\ U_{1,1} \end{vmatrix}, \quad A^1 = \begin{vmatrix} 0 & -c_0^2[1 + 3(1 + m_0)U_{1,1}] \\ -1 & 0 \end{vmatrix}, \tag{4.97}$$

where $c_0^2 = (\lambda+2\mu)\rho_0^{-1}$, $m_0 = 2(\nu_1+\nu_2+\nu_3)(\lambda+2\mu)^{-1}$. Here nonlinearity is related to the term a_{12} in $A^1 : -3c_0^2(1 + m_0)U_{1,1}$.

Finally *approximation* level means that the nonlinear terms appear in the evolution equation. Here this equation reads (Engelbrecht, 1983)

$$\frac{\partial v}{\partial \tau} + \frac{3}{2}\frac{1 + m_0}{\varepsilon c_0} v \frac{\partial v}{\partial \xi} = 0\,, \tag{4.98}$$

where ε is a small parameter. The second term in this equation is responsible for nonlinear effects. It is easy to follow the transformation of the coefficients which retain the simultaneous influence of the physical and geometrical nonlinearities.

There is an important feature to be pointed out. Quite usually, the evolution equations are solved in their dimensionless form. As a result, the coefficients of nonlinear terms from the basic level may be mixed up with coefficients related to other physical effects. Such a situation demonstrates clearly the deep interaction of various effects.

An example may be given from the theory of viscoelasticity where the simplest 1D model of the Voigt type material reads

$$\frac{\partial u}{\partial \tau} + u\frac{\partial u}{\partial \xi} = \Gamma^{-1}\frac{\partial^2 u}{\partial \xi^2}\,, \tag{4.99}$$

where all the dependent and independent variables are dimensionless (Engelbrecht, 1983) and the coefficient Γ is expressed by

$$\Gamma = \frac{3|1 + m_0|}{n_0}\tau_c u_0\,. \tag{4.100}$$

Here n_0 is the kinematic viscocity, m_0 the dimensionless parameter of nonlinearity (see above), τ_c is the wavelength of the input and u_0 its initial amplitude. Consequently, a certain ratio of nonlinear and viscous parameters is important rather than the parameters separately. In addition, the parameter is input-dependent. As a result, the inverse problem for determining the material properties on the basis of Burgers equation (4.99) is ill-posed (Engelbrecht, 1983). Equation (4.99) and its structure will be analysed in Section 5.1.3.

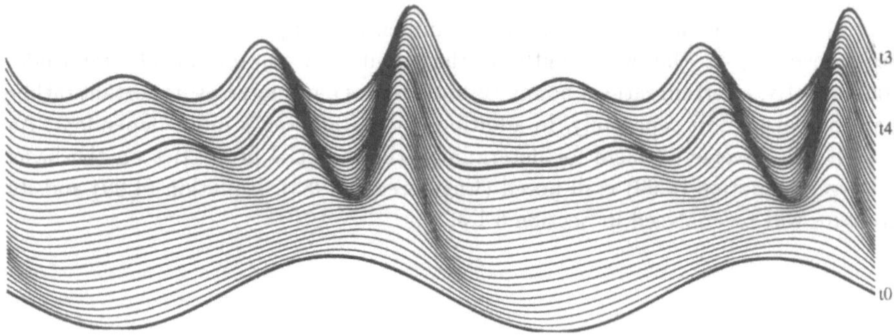

5. Wave phenomena: complexities in modelling

Beautiful are the things we see,
More beautiful those we understand,
Much the most beautiful those we do not comprehend.

Unknown author,
translated from Latin by A.Pais

5.1. COMPLICATED PHYSICAL PROPERTIES

5.1.1. Introductory remarks

From the very beginning of this book, we have been stressing the importance of being nonlinear. However, when dealing with real materials, this is not the only property to be taken into account. In general terms, there are two wide classes of effects, influencing wave propagation:

(i) space-dependent effects;
(ii) time-dependent effects.

Space-dependent effects are related to inhomogeneity, heterogeneity, internal structure, etc., while time-dependent effects are related to viscosity, relaxation, etc. Mathematical modelling of wave phenomena accounting for these effects must be certainly based on general principles of continuum mechanics. Special attention should be paid to the principle of equipresence, according to which the dependent variables (functionals) should depend on the same list of constitutive variables. In addition, all the effects of the same order should be taken into account. However, the order of a certain effect is à priori not explicitly known. As a result, the mathematical models dealing with so many effects apart from nonlinearity cannot be nothing else than extremely complicated in the very sense of the word, i.e. difficult to analyse, understand and explain. Thus, we accept a working notion—all the cases resp. models that besides nonlinearit-y/ies involve other effects, are considered to be complicated. We stress that this is a working notion and has nothing to do with strict rational analysis of continuum mechanics in terms of Truesdell and Noll (1965).

A logical question now follows whether it is possible to present an overwhelming treatment of all possible theories dealing with wave phenomena in media possessing complicated properties according to this working notion. In this book only rather general ideas are given and analysed for both classes of effects indicated above. Actually it means that the analysis will be centered around the essence of the axioms of neighbourhood and memory (Section 2.3), the first explaining space-dependent theories and the second—time-dependent theories. Further, when explaining space-dependent theories we rely upon Braun and Engelbrecht (1995).

5.1.2. Space-dependent theories

General description. The concept of homogeneous solids has successfully been applied to many technical problems. Modern physics and technology, however, cannot take the concept of homogeneity as the sole basis any more because of the growing need to use heterogeneous materials, short wavelengths, coupled fields, etc. Indeed, most materials are heterogeneous not only on the microscale but also on the meso- and macroscales due to the method of manufacture or formation. The latter concerns, for example, geomaterials. And modern technology is very much based on using high frequencies and effects of coupling between various fields. Much effort has been put into constructing the mathematical models of media that are not homogeneous. *Heterogeneous* media are composed as a superposition of several homogeneous media (Truesdell and Toupin, 1960). Inhomogeneity, in turn, means that the material properties vary in space due to the technology (casting, for example) or to prestress at a certain preliminary stage. Structured media are special heterogeneous media in which material properties vary according to some rule different for various scales. Examples are media with regular or stochastic distributions of voids, fibres, cracks, etc. To speak of a structured medium presumes that the properties of the medium are viewed in a scale large enough to recognize the inherent structure. While the properties of a medium are then different for every point of the material space, then in mathematical terms the concept of nonlocality might be of use. Physically it means to account for the dependence of stress and body forces at a given point on the fields in the whole body. Whether a certain effect of heterogeneity and structure means nonlocality must be explained for every case study. One should also stress that the modern non-equilibrium thermodynamics (rational thermodynamics, theory of irreversible processes) takes nonlocality, but in time rather than in space, i.e. the whole past history, into account. If so, then the question arises if and how the space nonlocality is shadowed by thermodynamics.

In physical terms, nonlocality means that a field variable at a point X at time t depends on other field variables at points of the body other than the point X at time t. In mathematical terms, the governing equations, as a rule, should contain functionals of the whole field rather than functions of its local value. The outcome then is in the form of an integro-differential equation. However, this is not always so, since in some cases integro-differential equations can be changed into the equivalent differential equations that may contain higher gradients explaining nonlocal effects.

Therefore we mainly keep to the physical explanation and deal with all mathematical models based on different interpretations. As said before, we look at nonlocalities in conjunction with wave propagation, motivated by the wide interest to nonlocal problems and results achieved so far.

The obvious reason for the nonlocality of field variables is the microstructure of a medium. Nonlocality is actually introduced in order to make a *homogeneous* model work for a *heterogeneous* material. This could be explained by a simple example, as follows.

Consider a block consisting of periodically alternating layers of two different elastic materials. The gross material is still elastic in the usual sense, but with elastic moduli varying in a discontinuous manner throughout the body. There is no nonlocal behaviour in the sense that the stress in one point is somehow related to the deformation occurring at a different point.

Now we are going to "homogenize" the material, i.e. we try to describe its gross behaviour by a constitutive equation involving only *constant* moduli. The simplest form is again an elastic material characterized by constant effective moduli. This crude homogenization is acceptable for static problems, but the dynamics of the model material is quite different from that of the original layered material. In order to improve the model we allow for nonlocal behaviour of the *model* material. Contrary to the fibre-type structure below there is no obvious mechanism for passing information between distant points of the layered material. Nonlocality is just the price that has to be paid for homogeneity.

The material may be granular, it may include dislocations, inclusions, voids and micro-cracks (Maugin, 1993). This is obviously the case of materials with dislocations (Kröner, 1968), metal-matrix composites (Taya and Arsenault, 1989), geomaterials displaying porosity and cracks (Nikolaevskij, 1990), etc. The crystal (lattice) structure has an impact on high-frequency (short wave-length) waves and is of importance in surface physics and high gradient fields (Maugin, 1994b). In liquid crystals (Oseen, 1933), the vectorial character of directors is of importance. Muscles, due to their fibre-type structure (myosin and actin filaments), also possess nonlocal properties (Colli, 1989). In all these cases it is clear that the stress at a point X depends on the deformation not only at X but also at other points Z of the body. If beside stress and strain also other fields, like electro-magnetic, temperature, etc., should be taken into account, the general picture becomes even more complicated (Maugin and Muschik, 1994).

The characteristic lengths in a material with microstructure may vary in a large range. For example, metal-matrix composites (Taya and Arsenault, 1989) are divided into three groups: dispersion-strengthened composites containing fine uniformly dispersed particles with diameter of $0.01 - 0.1\mu$m, particle-reinforced composites where the dispersed particles are of a diameter larger than 0.1μm, and fibre-reinforced composites with fibre lengths ranging from 0.1 to 250μm or even over the full length of the body. In geomaterials, the block structure in layers may have a characteristic length in km's that should be compared with wave-lengths two to three orders larger (Engelbrecht and Khamidullin, 1988). So the characteristic lengths of microstructures extend over the full range of micro-, meso- and macroscales.

Functional Nonlocality. In the shortest way we may express the essence of this theory by the statement that the stress at a point X_A is a functional of strains at all points of the body $\mathcal{B}$:

$$T_{iA}(X_K,t) = \underset{Z_K \in \mathcal{B}}{\mathcal{F}_{iA}}[x_{j,B}(Z_K,t); X_K]\,. \tag{5.1}$$

The additional X_K in the argument allows the functionals $\mathcal{F}_{iA}$ to depend explicitly on the material point at which it delivers the stress ("nonhomogeneous nonlocality"). A specific example may be given by following Edelen (1976). The free energy function

$$F = F(X_K, t, x_i, x_{i,A}, k_a) \tag{5.2}$$

must exist together with q functionals

$$k_a = \int_{\mathcal{D}} g_a[X_K, t, Z_K, \tau, x_i(Z_K,t), x_{i,A}(Z_K,\tau)]\mathrm{d}V(Z_K)d\tau \tag{5.3}$$

such that the stress becomes

$$T_{iA} = \frac{\partial F}{\partial x_{i,A}} + \int_{\mathcal{D}} \frac{\partial F}{\partial k_a}(Z_K,\tau)\frac{\partial g_a^*}{\partial x_{i,A}} dV(Z_K)\mathrm{d}\tau\,. \tag{5.4}$$

Here the q functions g_a^* are given by

$$\begin{aligned} g_a^*[X_K, t, Z_K, \tau; x_i(Z_K,t), x_{i,A}(Z_K,t)] &= \\ = g_a[Z_K, \tau, X_K, t; x_i(X_K,t), x_{i,A}(X_K,t)]\,, \end{aligned} \tag{5.5}$$

i.e. the arguments (X_K, t) and (Z_K, τ) have to be interchanged (Edelen, 1976). If F is independent of k_a, t and x_i, the constitutive equation describes an elastic body without any nonlocal behaviour, since the integral term in (5.4) vanishes. In a simple linear example, retaining nonlocality, the constitutive equation reduces to (Edelen, 1976)

$$T_{11}(X_1,t) = (\lambda + 2\mu)\left[U_{1,1}(X_1,t) + \int_{-\infty}^{+\infty} K(|\,X_1 - \xi\,|)U_{1,1}(\xi,t)\mathrm{d}\xi\right]\,, \tag{5.6}$$

where the kernel function $K(r)$ has to decrease rapidly enough for $r \to \infty$, to ensure the convergence of the integral.

Edelen (1976) has also dealt with the case of a nonlinear functional. The strain-energy function of a rod is assumed to be

$$F = \frac{1}{2}E[1 - f(k)]U_{1,1}^2\,, \quad f(k) \geq 0\,, \quad f(0) = 0 \tag{5.7}$$

with Young's modulus E and the functional

$$k = \int_0^L U_{1,1}^2(Z,t)\mathrm{d}Z\,. \tag{5.8}$$

This delivers the uniaxial stress

$$T_{11} = E\left(1 - f(k) - k\frac{\partial f}{\partial k}\right)U_{1,1} \tag{5.9}$$

still depending on the choice of the function f. The problem is to determine the value of k from given boundary data. If the rod is fixed at $X_1 = 0$ and a constant stress σ_0 is applied at the free end $X_1 = L$, then there will be a uniform strain

$$\varepsilon = \frac{\sigma_0}{E\left[1 - f(k) - k\frac{\mathrm{d}f}{\mathrm{d}k}\right]} \tag{5.10}$$

throughout the rod, where k is a real root of the nonlinear equation

$$k\left(1 - f(k) - k\frac{\mathrm{d}f}{\mathrm{d}k}\right)^2 = \frac{\sigma_0^2 L}{E^2}\,. \tag{5.11}$$

For further details, see Edelen (1976).

Returning to the three-dimensional case, *geometrical* nonlinearity is taken into account by using the full nonlinear expression of some appropriate strain tensor to generalize the $U_{1,1}$ in the one-dimensional case above. *Physical* nonlinearity means that the strain-energy functional depends also on terms of third or even higher orders in strain. The conventional theories are based on introducing just the third-order terms (Eringen, 1962), but for describing shape-memory alloys, for instance, one needs also the terms of fourth, fifth and sixth order in strain to be taken into account (Maugin and Cadet, 1991).

According to Maugin and Muschik (1994) nonlocality expressed by functionals is *true* nonlocality (c.f. weak nonlocality below).

Gradient Theory. The inclusion of higher gradients of displacement and velocity in the basic postulates means that generalized forces and generalized balance laws have to be formulated (Green and Rivlin, 1964). Second-grade materials are defined as materials for which stress is a function of both the deformation gradient $x_{i,A}$ and the second-order deformation gradient $x_{i,AB}$. In other words, the strain energy F should be determined by a function

$$F = F(x_{i,A}; x_{i,AB}; \ldots)\,. \tag{5.12}$$

Beside the earlier papers from the period of establishing the theory (Lardner, 1969; Mindlin and Eshel, 1968), there is a new trend to use gradient theory (Erofeyev and Potapov, 1993). In some sense gradient theory accounts for *weak* nonlocality (Maugin and Muschik, 1994). The gradient theories may be derived from the functional approach via series expansion. An example will be presented later.

In gradient theories, the stress depends on just the same deformation gradients of first and higher orders as the strain-energy function. But, what is more important, each of the higher-order deformation gradients included gives rise to a corresponding higher-order stress tensor. In a second-grade material, following Mindlin and Eshel (1968), we have to distinguish between the stress

$$\tau_{i,A} = \frac{\partial F}{\partial x_{i,A}}\,, \tag{5.13}$$

which is identical to the usual Piola-Kirchhoff stress T_{iA}, and the "double" stress

$$\mu_{iAB} = \frac{\partial F}{\partial x_{i,AB}}\,. \tag{5.14}$$

One should, in principle, continue structuring the derivatives of F, if (5.12) involves even higher gradients.

Instead of the deformation gradients $x_{i,A}, x_{i,AB}, \ldots$ the corresponding displacement gradients of the first and higher orders $U_{i,A}, U_{i,AB}, \ldots$ can also be used. These are better suited for series expansions. Retaining only linear terms the stress in a one-dimensional case could be written as

$$T_{11} = (\lambda + 2\mu)U_{1,1} + \kappa U_{1,111} + \ldots \tag{5.15}$$

where λ, μ are the usual Lamé parameters and κ denotes a constant involving the coefficients of the series expansion of (5.12). Here also, like in functional nonlinearity, the nonlinearity enters via terms of higher orders in $U_{i,A}$ in the free energy function F. Just as an example, instead of (5.15) we might consider

$$T_{11} = (\lambda + 2\mu)U_{1,1} + \left(\tfrac{1}{2}\lambda + \mu + 3\nu_1 + 3\nu_2 + 3\nu_3\right) U_{1,1}^2 + \kappa U_{1,111} + \dots \qquad (5.16)$$

where the third-order elastic moduli ν_i have been introduced earlier.

The expression (5.15) should be compared with (5.6) that consists of an integral term accounting for nonlocality, while in (5.15) the third-order derivative $U_{1,111}$ does the same. However, in (5.15) nonlocality is weak, according to Maugin and Muschik (1994). Intuitively, one could think about discretizing the derivatives in (5.15) to understand the difference between local and nonlocal behaviour. While the first-order derivative needs knowing the values of U_1 at only two neighbouring nodes, then the third-order derivative takes at least the values at four nodes into account. This means that the dependence domain of (5.15) is considerably larger than that of the "local" case $\kappa = 0$.

Internal Variables. An excellent presentation of the fundamental aspects of continuum mechanics with internal variables is presented by Maugin and Muschik (1994) with quite a lot of applications given in the second part of the paper. Here we follow their general ideas.

The dependent variables σ like stress T_{iA} will be functions of both the observable variables χ and the internal variables α (see Section 2.4)

$$\sigma = \sigma(\chi, \alpha)\,. \qquad (5.17)$$

The internal variables α are governed by their own equations, usually of kinetic type

$$\dot{\alpha} = f(\chi, \alpha)\,. \qquad (5.18)$$

Both equations are part of the constitutive law describing the behaviour of the material and have to comply with the general rules of continuum mechanics, such as the principle of material frame indifference. The choice of an internal variable depends upon the level of observation. Actually, the α's compensate for our lack of precise description of the internal mechanisms governing the gross behaviour of a medium.

Equations (5.17), (5.18) actually reflect *temporal* nonlocality. In principle, the internal variable α could be expressed as a functional of $\chi(t)$ by integration of (5.18), thus allowing the dependent variable σ to depend on the total history of state. But in spatial sense the theory is still local. To allow for *spatial* nonlocality, the constitutive law should depend at least on the gradients $\alpha_{,A}$ of the internal variables. Suppose the strain energy per unit volume is given by a function $W(\chi, \alpha, \alpha_{,A})$. The internal variables are accompanied by an extra entropy flux, which in many, though not all, cases may be represented by a dissipation potential $\mathcal{D}(\chi, \alpha, \alpha_{,A}, \dot{\alpha})$ homogeneous of degree two in $\dot{\alpha}$. It can then be inferred from the Clausius-Duhem inequality that

$$\frac{\delta W}{\delta \alpha} + \frac{\partial \mathcal{D}}{\partial \dot{\alpha}} = 0 \qquad (5.19)$$

holds, where $\delta/\delta\alpha$ denotes the Euler-Lagrange derivative. Working this out in full length yields

$$\frac{\partial W}{\partial \alpha} - \frac{\partial^2 W}{\partial \chi \partial \alpha_{,A}} \chi_{,A} - \frac{\partial^2 W}{\partial \alpha \partial \alpha_{,A}} \alpha_{,A} - \frac{\partial^2 W}{\partial \alpha_{,A} \partial \alpha_{,B}} \alpha_{,AB} + \frac{\partial \mathcal{D}}{\partial \dot{\alpha}} = 0\,. \tag{5.20}$$

The last term is homogeneous of degree one in $\dot{\alpha}$. If (5.20) is solved for $\dot{\alpha}$, one obtains an equation of the form

$$\dot{\alpha} = f(\chi, \chi_{,A}; \alpha, \alpha_{,A}, \alpha_{,AB}) \tag{5.21}$$

which now describes the internal kinetics. If the dependence of the strain-energy function on the gradient $\alpha_{,A}$ is abandoned, the Euler-Lagrange derivative in (5.19) is reduced to the partial derivative, all the second partial derivatives in (5.20) disappear, and the kinetic equation (5.21) assumes the simple form (5.18) valid in the local case.

Internal-degree-of-freedom interpretation (polar theories). As mentioned before, internal degrees of freedom give rise to additional dynamical equations. Most typical examples are micropolar media, where the balance of angular momentum retains its full dynamical form involving rotary inertia.

In general, the material points of a medium are considered to possess properties similar to rigid or deformable particles. These properties are introduced by prescribing every point of a medium through a set of vectors (directors), which means adding independent degrees of freedom to every point. If the directors form a rigid triad, the medium is micropolar, if the triad is deformable, the medium is micromorphic (Eringen and Kafadar, 1976). In addition to the displacement, every point (particle) has then also a rotational degree of freedom described by a vector ϕ_r that enters to the constitutive equations which, therefore, involve additional elastic constants. Moreover, both the balance of momentum and the balance of moment of momentum include inertial terms. For example, the coupled constitutive equations of a linear polar elastic body are

$$\begin{aligned} t_{ij} &= A_{ijkl}(U_{k,l} + \varepsilon_{krl}\phi_r) + B_{ijkl}\phi_{k,l}\,, \\ m_{ij} &= B_{ijkl}(U_{k,l} + \varepsilon_{krl}\phi_r) + C_{ijkl}\phi_{k,l}\,, \end{aligned} \tag{5.22}$$

where $A_{ijkl}, B_{ijkl}, C_{ijkl}$ are elastic constants and ε_{krl} denotes the permutation symbol. (The corresponding equations $(1.6.16)_{1,2}$ in Eringen and Kafadar (1976) are not correct).

The theories of polar continua go back to E. and F. Cosserat and a comprehensive review on this topic is given by Eringen and Kafadar (1976) (see also Kunin, 1982; 1983). As said before, this interpretation is closely related to the internal-variable approach (see above).

Eshelby Stress Formalism. Eigenstrains, according to Mura (1993) are the strains of non-elastic origin such as thermal expansion, phase transformation, plastic yield, etc. If a constant, non-zero eigenstrain e^*_{kl} is confined to a domain Ω of a homogeneous elastic body B, then Ω is called an inclusion. The total strain e_{ij} in an arbitrary point x_r inside or outside the inclusion can be expressed by

$$e_{ij} = S_{ijkl}(x_r)e^*_{kl}\,, \tag{5.23}$$

where S_{ijkl} is the Eshelby tensor (Mura, 1991). Closed-form solutions are known for some regular shapes of inclusions embedded in an infinite isotropic elastic body. Actually, Eshelby's tensor is a measure of the mechanical interaction of an isolated defect with its elastic surrounding.

The theory is directly applicable to inhomogeneous bodies provided the volume fraction of the inhomogeneities is sufficiently low, so that interactions between them are negligible. If Ω now denotes the union of all inclusions or inhomogeneities, the total strain energy of the body B is

$$W = \frac{1}{2}\int_{\mathcal{D}} T^0_{ij}e^0_{ij}\mathrm{d}V + \frac{1}{2}\int_{\Omega} T^0_{ij}e^*_{ij}\mathrm{d}V\,, \tag{5.24}$$

where the subscript "0" marks the applied field. From this equation the effective elastic moduli can be calculated (Taya and Arsenault, 1989).

The Eshelby tensor S_{ijkl} must not be confused with Eshelby's stress tensor. Eshelby (1951) has introduced it under the name Maxwell tensor of elasticity and its dynamic generalization as the energy-momentum tensor. Substantial progress in the interpretation, generalization and application of Eshelby's concept has still been made within the recent years (Maugin, 1992a; Epstein, 1993; Kröner, 1993). Contemporary theory uses the Eshelby's energy-momentum tensor with a real ingenuity. It has been shown by Maugin (1992a) that this tensor can be interpreted as the material tensorial flux associated with the conservation of the so-called pseudomomentum. As a result, keeping his notations, the general equation of motion reads

$$\left.\frac{\partial}{\partial t}\mathcal{P}\right|_{X\,\mathrm{fixed}} - \mathrm{div}_R\,\boldsymbol{b} = \boldsymbol{f}^{\mathrm{inh}} \tag{5.25}$$

where $\mathcal{P}$ is the pseudomomentum material vector, $\mathbf{b}$ is the dynamical version of Eshelby's stress tensor and $\mathbf{f}^{\mathrm{inh}}$ is the material force accounting for elastic inhomogeneities and body forces. Such an approach has served to explain grade-two elasticity, solitonic effects, dynamical fracture, etc. (Maugin, 1992a; 1995).

Mathematical Models and Evolution Equations. The polar and gradient theories yield equations with higher-order derivatives. The simplest case of a polar material with scale factor ℓ_0, advocated by Kunin (1982; 1983), leads to none other than the Korteweg-de-Vries equation

$$u_\tau + uu_\xi + a_3 u_{\xi\xi\xi} = 0\,, \quad a_3 = \mathrm{const.} \tag{5.26}$$

Here and further, the proper scaling is taken to simplify the evolution equations as much as possible (Engelbrecht, 1983) and indices denote the differentiation. Equation (5.26) as the simplest evolution equation capable to model microstructure, certainly describes only weak nonlocality (Maugin and Muschik, 1994). A similar result is obtained in the first approximation for waves in waveguides (rods) to model the influence of boundaries (geometric dispersion) (Engelbrecht, 1983) while more detailed studies show the need to use a more complicated wave equation (Samsonov,

1994) where the splitting-up to single (unidirectional) waves is impossible (see Section 5.2). The gradient theory shows the same Eq. (5.26) as shown by Erofeyev and Potapov (1993).

The most remarkable property of Eq. (5.26) is the existence of localized solitary waves—solitons (see Section 5.3).

If higher-order gradients are included then the localization is weaker. Boyd (1990) has shown the existence of "nonlocal solitons" or "nanopterons" for the equation

$$u_\tau + uu_\xi + a_3u_{\xi\xi\xi} + a_5u_{5\xi} = 0\,, \quad a_3 = \text{const.}\,, \quad a_5 = \text{const.} \tag{5.27}$$

These nonlocal solitons have beside their core resembling the soliton of Eq. (5.26) also the wings attached to both sides.

Using a one-dimensional simplified approach of an internal-degree-of-freedom interpretation, Nikolaevskij (1990) has derived an evolution equation with terms up to the sixth order

$$u_\tau + uu_\xi + a_2u_{\xi\xi} + a_3u_{\xi\xi\xi} + a_4u_{4\xi} + a_5u_{5\xi} + a_6u_{6\xi} = 0\,, \tag{5.28}$$

where a_i, $i = 2, 3, \ldots 6$ are constants. This equation models both the dispersion and the attenuation/amplification and was derived for geomaterials. In this context, similar results obtained with many higher-order derivatives in the evolution equation, are derived by Kawahara and Toh (1988) and Normand et al. (1977).

Up to now, the nonlinearity was accounted for only in its very first approximation —the quadratic potential $\frac{1}{2}v^2$ resulting in the vv_ξ term. Maugin (1995) has noted the equation

$$u_\tau + nuu_\xi + m(uu_\xi)_{\xi\xi} + a_3u_{\xi\xi\xi} + a_5u_{5\xi} = 0 \tag{5.29}$$

with constant coefficients. Preliminary analysis has shown the significance of a quartic potential $\frac{1}{2}v^2 + \frac{1}{4}kv^4$, k = const, with two minima resulting in different solitary waves that need stability analysis.

Within the non-integrable evolution equations (5.28) and (5.29) the evolution equation with a right-hand side of lower-order terms should be noted.

Such an equation was derived by Engelbrecht and Khamidullin (1988) accounting for cubic dispersion, quadratic nonlinearity, and an additional forcing. It has the form

$$u_\tau + uu_\xi + a_3u_{\xi\xi\xi} = \varepsilon f(u) \tag{5.30}$$

with $f(u) = u - b_2u^2 + b_3u^3$ while ε is a small parameter to stress the weak influence of the forcing. The model permits amplification of a propagating wave in a certain amplitude domain $u \in [u_1, u_2]$, $u_2 > u_1$, where u_1, u_2 are the real positive roots of $f(u)$. For smaller waves $(u < u_1)$ attenuation takes place. Characteristically, the amplified stable solitary wave is asymmetric (Engelbrecht, 1991).

Another important class of mathematical models is based on integro-differential equations. These "true" mathematical models are of a great importance in contemporary mathematical physics. An excellent review on the properties of solutions of nonlinear nonlocal integro-differential equations is presented by Naumkin and Shishmarev (1994), where the examples are drawn from all branches of physics. Here we concentrate more on those equations that result from theories enlisted above.

Whitham (1974) combines the nonlinearity and dispersion on the level of the evolution equation simply by physical considerations yielding

$$u_\tau + uu_\xi + \int_{-\infty}^{+\infty} K(\xi - \eta) u_\eta(\eta, t) \mathrm{d}\eta = 0 \,. \tag{5.31}$$

If the kernel is taken as $K(x) = c_2\delta''(x)$ where $\delta(x)$ is the Dirac function, then the Korteweg-de-Vries equation (5.26) follows from (5.31). For shallow-water waves where nonlocality is emphasized by the influence of boundaries, the kernel is (Whitham, 1974)

$$K_s(x) = \frac{1}{2\pi} \int_{-\infty}^{+\infty} \left(\frac{g}{\kappa} \tanh \kappa h_0 \right)^{\frac{1}{2}} \exp(i\kappa x) \mathrm{d}x \,, \tag{5.32}$$

where g stands for the acceleration, h_0 for the depth, and κ denotes the wave number. A useful approximation to (5.32) is

$$K_0(x) = \mu e^{-\nu |x|} \,, \tag{5.33}$$

with constants determined from the behaviour at $x \to \infty$ resulting in $\mu = \pi/4$, $\nu = \pi/2$. This kernel is able to describe the peaking of symmetric solitary waves as well as the breaking of asymmetric waves.

Eleonsky et al. (1994) have analysed Eq. (5.31) with rapidly decreasing oscillating kernel of the type

$$K(x) = G(\sigma) e^{-x/\sigma} \cos x \tag{5.34}$$

where σ is a constant. The solution for this case is obtained by splitting the governing equation into a system of equations involving the initial nonlinear field and auxiliary fields interacting locally with it. The solution looses its smoothness and is characterized by noncontinuability of the solution causing the peaking of a localized solution.

For one-dimensional crystalline bodies Eringen (1992) has used the kernel

$$K(x) = \begin{cases} \dfrac{1}{a}\left(1 - \dfrac{|x|}{a}\right), & |x| \leq a \,, \\ 0, & |x| > a \,, \end{cases} \tag{5.35}$$

predicted by the lattice dynamics (a is the lattice parameter). The convex dispersion curve obtained from this model is rather close to that resulting from the Born-Kármán model. A suitable approximation of K in the Fourier space

$$K_F = (1 + \varepsilon^2\kappa^2 + \gamma^4\kappa^4)^{-1} \,, \quad \varepsilon = \mathrm{const} \,, \quad \gamma = \mathrm{const} \,, \tag{5.36}$$

gives the wave equation

$$c_0^2 U_{XX} - U_{tt} + \varepsilon^2 U_{XXtt} - \gamma U_{4Xtt} = 0 \,, \quad c_0 = \mathrm{const.} \tag{5.37}$$

The evolution equation, when nonlinearity is added according to Whitham (1974) (c.f. Eq. (5.31)), is

$$u_\tau + uu_\xi + a_3 u_{\xi\xi\xi} + a_5 u_{5\xi} = 0 \,, \tag{5.38}$$

i.e. the same as obtained for an example from the gradient theory (c.f. Eq. (5.27)).

In quantum mechanics, a model theory of gravitating particles with nonlinear interaction leads to the nonlinear Schrödinger equation (Efinger, 1984)

$$i\phi_t + \Delta\phi + \phi \int |\phi(y,t)|^2 K(x - y) \mathrm{d}y = 0 \tag{5.39}$$

where ϕ is a space-time function.

Last but not least, the kinetic equations for internal variables should be mentioned. There is no general theory yet but some examples demonstrate the main idea. Maugin and Muschik (1994) argue about the equation

$$\alpha_t = h(v, D, \alpha, \nabla\alpha, \theta, \nabla\theta, \rho)\,, \tag{5.40}$$

where D is the strain rate, θ is the temperature and ρ is the density. Among the examples a quartic potential W and a quadratic dissipation potential $\mathcal{D}$ (c.f. Eq. (5.19)) give

$$\tau\alpha_t = -(a\alpha + b\alpha^3) + c\alpha_{XX}\,, \tag{5.41}$$

where τ is the relaxation time and a, b, c are constants. Equation (5.41) is used for describing phase transitions. Actually, if W has several extrema then a factorization may be useful resulting in

$$\tau\alpha_t = -k\prod_j(\alpha - \alpha_j) \tag{5.42}$$

where the α_j's are local extrema (Maugin and Muschik, 1994). The role of Eqs. (5.41), (5.42) in wave propagation needs more studies.

5.1.3. Time-dependent theories

General description. Elastic deformation is reversible and time-independent. Contrary to that, the irreversible and time-dependent deformation is called viscous. Metals and plastics (polymers) are usually viscoelastic, i.e. they possess mixed properties. The symptoms of viscoelasticity are different (Ferry, 1961) as demonstrated by simple examples. A viscoelastic body under constant stress deforms slowly in time and this is called creep. The same body constrained at constant deformation needs less stress in time to keep this deformation. That is called relaxation. When such a body is subjected to oscillating stress then the strain will not be exactly in phase with the stress.

Using these simple considerations, some more complicated situations may easily be described. Suppose a stress is suddenly applied to a viscoelastic body. This induces an instantaneous deformation which is then followed by creep as said above. When two stress states are applied nonsimultaneously (i.e. superimposed upon each other) then the situation is the following. The first excitation causes, as above, the instantaneous response after which the time-dependent process (creep) follows. When the second excitation is in effect then the body reacts simultaneously but continues at the same time to creep as the aftermath to the first excitation. This is actually the memory effect—the body "remembers" the first excitation (Christensen, 1971). Let us suppose now that a stress is applied very slowly. Then it is logical to expect a different behaviour from the situation described above. This is typical for polymer materials as shown by Nunziato et al. (1974)

Consequently, one should distinguish between instantaneous (i.e. fast response) and equilibrium (i.e. slow response) stress-strain relations that means a pronounced influence of the rate of loading. If nonlinearity is taken into account then the situation becomes evidently more complicated.

The reason for viscous effects briefly described above is explained in terms of the thermal energy possessed by atoms and molecules in solids (Suh and Turner,

1975). In other words, the mechanical work done during the deformation of a viscoelastic body, is dissipated in the form of thermal energy. It is then obvious that viscosity depends largely on temperature of the solid. From the structural side, viscosity is controlled by several basic properties, such as the size of the molecule, the molecular structure and the shape, and the bond strength (Suh and Turner, 1975). This has given rise to the molecular viscoelastic theory of polymers (Han, 1976). For our purpose, however, we are here more interested in the phenomenological theory (Flügge, 1967; Christensen, 1971) and present below the basic stress-strain relationships.

Quite a common representation for 1D models are the spring-dashpot structural models that give an excellent idea about the possible “slow” and “fast” deformation of an imaginary element.

Here we base our considerations on the expressions presented in Chapters 2 and 3 in 1D case and rectangular coordinates in order to get more transparent and shorter description.

The simplest (and probably the most widely used) viscoelastic body is that of Voigt (or Kelvin). Using Eq. (2.17) we have

$$T_{11} = (\lambda + 2\mu)U_{1,1} + \left(\tfrac{1}{2}\lambda + \mu + 3\nu_1 + 3\nu_2 + 3\nu_3\right) U_{1,1}^2 + \left(\xi_\nu + \tfrac{4}{3}\eta_\nu\right) \dot{U}_{1,1} \,. \quad (5.43)$$

Higher order terms in $\dot{U}_{1,1}$ could be easily imagined (Truesdell and Noll, 1965) but the physical interpretation of their coefficients could raise the problems. The stress-strain relation (5.43) is of the differential type.

Another widely used model is the so-called standard viscoelastic body which is the simplest case of the memory-type (integral-type) material. The kernel G^{KLIJ} in Eq. (2.19) is assumed of the exponential type, i.e. possessing the property of the fading memory

$$G^{KLIJ} \sim \exp\left(-\frac{t-\tau}{\tau_o}\right) H(t-\tau)\,, \quad (5.44)$$

where τ_0 is the relaxation time and $H(t)$ the Heaviside function. There are theories using a series representation of the kernel function that physically means the existence of a spectrum of relaxation times τ_i instead of a single relaxation time $\tau = \tau_1$ (Nunziato et al., 1974).

Now, keeping the physical and geometrical nonlinearities for the reversible part of the stress tensor and linear approximation for the time-dependent part, we obtain

$$T_{11} = (\lambda + 2\mu)U_{1,1} + \left(\tfrac{1}{2}\lambda + \mu + 3\nu_1 + 3\nu_2 + 3\nu_3\right) U_{1,1}^2 + \\ +\varepsilon_1(\lambda + 2\mu) \int_0^t \frac{\partial U_{1,1}}{\partial \tau} \exp\left(-\frac{t-\tau}{\tau_0}\right) \mathrm{d}\tau\,, \quad (5.45)$$

where ε_1 is a dimensionless constant. Integrating per part, another form is easily deduced

$$T_{11} = (\lambda + 2\mu)(1+\varepsilon_1)U_{1,1} + \left(\tfrac{1}{2}\lambda + \mu + 3\nu_1 + 3\nu_2 + 3\nu_3\right)(1+\varepsilon_1)U_{1,1}^2 - \\ -(1+\varepsilon_1)(\lambda + 2\mu)\tau_0^{-1} \int_0^t U_{1,1}(\tau) \exp\left(-\frac{t-\tau}{\tau_0}\right) \mathrm{d}\tau\,. \quad (5.46)$$

Mathematical Models and Evolution Equations. The Voigt model (5.43) yields in properly chosen dimensionless variables (Engelbrecht, 1983)

$$u_\tau + auu_\xi = \Gamma^{-1} u_{\xi\xi} \tag{5.47}$$

that is the Burgers equation. This evolution equation is widely known in fluid dynamics and nonlinear acoustics (Naugolnykh and Ostrovsky, 1990). The parameter Γ is known as the acoustic Reynolds number and is determined by the expression

$$\Gamma = 3 \mid 1 + m_0 \mid \tau_c u_0 n_0^{-1} , \tag{5.48}$$

where the notations are explained earlier (Section 4.4). Generally speaking, parameter Γ expresses similarity. If $\Gamma \gg 1$, nonlinear effects dominate; if $\Gamma \ll 1$, dissipative effects dominate. The structure of parameter Γ shows that it is the product of two parameters

$$\Gamma = G I_0 , \tag{5.49}$$

$$G = 3 \mid 1 + m_0 \mid n_0^{-1} , \quad I_0 = \tau_c u_0 . \tag{5.50}$$

Parameter G (s cm^{-2}) is the material parameter representing the influence of nonlinearity and viscosity. Parameter I_0 (s^{-1} cm^2) characterizes the input. Thus for a given medium the acoustic Reynolds number depends essentially on the parameters of the input. In acoustic problems usually Re$\gg 1$, $u_0 c_0^{-1} = M \ll 1$, but there is a wide range of possible values for Γ. In shock wave theory M may have a greater value.

The Burgers equation is the simplest for describing simultaneously nonlinear and dissipative effects, and it is difficult to overestimate its importance in the theory of dissipative processes (Lighthill, 1956). Its typical solution is the shock profile (Whitham, 1974).

For a medium of the memory type (5.45) the situation is more complicated and we start writing up the mathematical models in their natural form before transforming them to evolution equations.

The equations of motion are easily obtained either on the basis of (5.45) as

$$(\lambda + 2\mu)[1 + 3(1 + m_0)U_{1,1}]U_{1,11} + \\ +\varepsilon_1(\lambda + 2\mu)\left[\int_0^t \frac{\partial U_{1,1}}{\partial \tau} \exp\left(-\frac{t-\tau}{\tau_0}\right) d\tau\right]_{,1} - \rho_0 \ddot{U}_1 = 0 \tag{5.51}$$

or on the basis of (5.46) as

$$(\lambda + 2\mu)(1 + \varepsilon_1)[1 + 3(1 + m_0)U_{1,1}]U_{1,11} - \\ -\varepsilon_1(\lambda + 2\mu)\tau_0^{-1}\left[\int_0^t U_{1,1} \exp\left(-\frac{t-\tau}{\tau_0}\right) d\tau\right]_{,1} - \rho_0 \ddot{U}_1 = 0 . \tag{5.52}$$

It is obvious that (5.51) is based on the equilibrium velocity $c_e = [(\lambda + 2\mu)\rho_0^{-1}]^{1/2}$ and (5.52) on the instantaneous velocity $c_i = [(1 + \varepsilon_1)(\lambda + 2\mu)\rho_0^{-1}]^{1/2}$. From thermodynamics $\varepsilon_1 > 0$, hence $c_i > c_e$. In fluids ε_1 is rather small, of the order of 10^{-5} (Naugolnykh and Ostrovsky, 1990), but in polymers, for example, ε_1 has larger values ($\sim 10^{-2}$) (Nunziato et al., 1974). There exists still a third possibility of presenting the equation of motion on the basis of the respective differential relation rather than of (5.45) or (5.46). In this case we get

$$\tau_0 \frac{\partial}{\partial t}[(\lambda + 2\mu)(1 + 3(1 + m_0)U_{1,1})U_{1,11} - \rho_0 \ddot{U}_1] +$$

$$+(\lambda+2\mu)[1+3(1+m_0)U_{1,1}]U_{1,11}-\rho_0\ddot{U}_1+$$
$$+\varepsilon_1(\lambda+2\mu)\tau_0\dot{U}_{1,11}=0\,. \tag{5.53}$$

There is now the need to distinguish between equilibrium and instantaneous processes introducing the ray coordinates either by

$$\xi_e \sim c_e t - x\,, \quad \tau_e \sim \varepsilon x \tag{5.54}$$

or by

$$\xi_i \sim c_i t - x\,, \quad \tau_i \sim \varepsilon x\,. \tag{5.55}$$

The evolution equations are then obtained, respectively, in the following form

$$\frac{\partial u}{\partial \tau_e} + \mathrm{sign}(1+m_0)u\frac{\partial u}{\partial \xi_e} - \Xi_c\frac{\partial}{\partial \xi_e}\int_0^{\zeta_e}\frac{\partial u(z)}{\partial z}\exp\left(-\frac{\xi_e - z}{Z_e}\right)\mathrm{d}z = 0\,, \tag{5.56}$$

$$\frac{\partial u}{\partial \tau_i} + \mathrm{sign}(1+m_0)u\frac{\partial u}{\partial \xi_i} +$$
$$+\Xi_i Z_i^{-1}(1+\varepsilon_1)^{-1}\frac{\partial}{\partial \xi_i}\int_0^{\zeta_i}u(z)\exp\left(-\frac{\xi_i - z}{Z_i}\right)\mathrm{d}z = 0 \tag{5.57}$$

and

$$Z_e\frac{\partial}{\partial \xi_e}\left(\frac{\partial u}{\partial \tau_e} + \mathrm{sign}(1+m_0)u\frac{\partial u}{\partial \xi_e}\right) +$$
$$+\frac{\partial u}{\partial \tau_e} + \mathrm{sign}(1+m_0)u\frac{\partial u}{\partial \xi_e} - \Xi_e Z_e\frac{\partial^2 u}{\partial \xi_e^2} = 0\,. \tag{5.58}$$

The dimensionless parameters Ξ_k, Z_k, $k = i, e$, are fundamental in order to estimate the main properties of a process. They are given by

$$\Xi_k = \frac{1}{3}\varepsilon_1 c_k(\mid 1+m_0 \mid u_0)^{-1}\,, \tag{5.59}$$

$$Z_k = \tau_0 c_k \tau_c^{-1} \tag{5.60}$$

and satisfy

$$\Xi_i = \Xi_e(1+\varepsilon_1)^{1/2}\,, \quad Z_i = Z_e(1+\varepsilon_1)^{1/2}\,. \tag{5.61}$$

If $\Xi_k \gg 1$, the nonlinear effects are small; if $\Xi_k \ll 1$, the nonlinear effects dominate. The parameter Z_k includes the scale length $l_0 = \tau_0 c_k$, i.e. the distance covered by a wave during the relaxation time. Relation of scale length l_0 to wave length τ_c estimates the influence of the dispersion. If $Z_k \ll 1$, the dispersive effects are small; if $Z_k \gg 1$, the group-velocity change must be at a maximum. The greatest influence of dispersion must be at $Z_k \sim 1$. It is understandable that the same results follow from direct analysis of the dispersive relation obtained on the basis of Eq. (5.51). The full dispersion relation is

$$\omega = \pm c_e k[1 + \varepsilon_1\omega^2\tau_0^2(1+\omega^2\tau_0^2)^{-1} + i\varepsilon_1\omega\tau_0(1+\omega^2\tau_0^2)^{-1}]^{1/2}\,. \tag{5.62}$$

If $\omega\tau_0 \gg 1$, then

$$\omega = \pm c_e k[1+\varepsilon_1 + i\varepsilon_1(\omega\tau_0)^{-1}]^{1/2}\,, \tag{5.63}$$

which corresponds to the case $Z_e \gg 1$. If $\omega\tau_0 \ll 1$, then

$$\omega = \pm c_e k(1+i\varepsilon_1\omega\tau_0)^{1/2}\,, \tag{5.64}$$

which corresponds to the case $Z_e \ll 1$. The approximate evolution equations can then be constructed on the basis of estimated Z_k and the dispersion relations (5.63) and (5.64) are used as additional estimations. For that the reader is referred to Engelbrecht (1983).

5.1.4. Time- and space-dependent theories

There are but a few data about the simultaneous effects of time and space nonlocalities. A model evolution equation in the same framework as adopted above can be (Maugin, 1979)

$$u_\tau + uu_\xi + \int_{-\infty}^{t} \int_{-\infty}^{+\infty} K(\xi - \eta, \tau - s) u_\eta(\eta, s) \mathrm{d}\eta \mathrm{d}s = 0 . \tag{5.65}$$

For practical applications the kernel $K(\xi, \tau)$ is usually split up into various parts. As an example, we construct an integro-differential counterpart to the Korteweg-de-Vries-Burgers equation

$$u_\tau + uu_\xi - a_2 u_{\xi\xi} + a_3 u_{\xi\xi\xi} = 0 , \tag{5.66}$$

when the third term ($a_2 > 0$) expresses dissipation (nonlocal time effect) and the fourth term dispersion (nonlocal space effect). Following Whitham (1974), the fourth term could be generalized to

$$\int_{-\infty}^{+\infty} K_1(\xi - \eta) u_\eta \mathrm{d}\eta . \tag{5.67}$$

Time effects could be generalized in the same way (Nunziato et al., 1974). In terms of stress, the third term in (5.66) stems from the dissipative part of the stress tensor

$$T_{11}^{D} = \int_{-\infty}^{t} \bar{K}(t - t') \frac{\partial U_{1,1}(X, t')}{\partial t'} \mathrm{d}t' , \tag{5.68}$$

where t, t' have real time dimensions. In the moving frame the divergence term $T_{11,1}$ gives rise to an expression of the form

$$-\frac{\partial}{\partial \xi} \int_{-\infty}^{\xi} K_2(\xi - \eta) u_\eta \mathrm{d}\eta , \tag{5.69}$$

because the independent variable ξ involves both t and X. The full equation reads then

$$u_\tau + uu_\xi + \frac{\partial}{\partial \xi} \int_{-\infty}^{\xi} K_2(\xi - \eta) u_\eta \mathrm{d}\eta + \int_{-\infty}^{+\infty} K_1(\xi - \eta) u_\eta \mathrm{d}\eta = 0 \tag{5.70}$$

with two different kernels K_1 and K_2. By no means is Eq. (5.70) equivalent to Eq. (5.66).

For a relaxing medium of the standard viscoelastic type exponential kernels are widely used. In this case, with $K_1(\xi) = 0$ and $K_2(\xi) = a \exp(-k\xi)$, Eq. (5.70) is transferable to a differential equation with higher-order derivatives describing wave hierarchy (Engelbrecht, 1983).

A more general approach was proposed by Sugimoto and Kakutani (1985) using a kernel of the form

$$K_2(\xi) = \frac{a_{1+\nu} \xi^\nu}{\Gamma(1 - \nu)} , \quad 0 < \nu < 1 , \tag{5.71}$$

where $a_{1+\nu}$ is some constant and $\Gamma(\nu)$ denotes the Gamma-function. By using the definition of fractional derivative

$$\frac{\partial^\nu(\ldots)}{\partial t^\nu} = \frac{1}{\Gamma(1-\nu)} \int_{-\infty}^{t} \frac{1}{(t-\tau)^\nu} \cdot \frac{\partial(\ldots)}{\partial \tau} \mathrm{d}\tau \tag{5.72}$$

it is possible to obtain from Eq. (5.70) with $K_1(\xi) = 0$ a more general evolution equation of a pseudo-differential type

$$u_\tau + uu_\xi - a_{1+\nu} \frac{\partial^{1+\nu} u}{\partial \xi^{1+\nu}} = 0 \tag{5.73}$$

that interpolates between parabolic Burgers type ($\nu = 1$) and simple nonlinear wave type ($\nu = 0$). The solution of Eq. (5.73) demonstrates scaled after-effects (Sugimoto and Kakutani, 1985).

5.2. WAVES IN COMPLICATED STRUCTURES

5.2.1. Direct influence of geometry

Until now we have been speaking about a 3D medium or its simplification to a 1D case. Clearly there is no influence of geometry on the macroscale. In bounded media, however, the boundaries can be treated as certain constraints and this influence should be certainly taken into account.

A 1D medium is a very rough simplification for a 1D object like a rod. The real wave process in this situation is much more complicated and can be briefly characterized as wave phenomena in waveguides (Miklowitz, 1980). The most important effect is due to the presence of a certain characteristic length based on the geometry of a waveguide. This leads immediately to the dispersion, i.e. the wave velocity depends on wavelength (see Chapter 1).

A naive model for a 1D waveguide may be built up using the continuum theory in 1D with the so-called Rayleigh correction (Tanaka and Motoyama, 1974). In the case of a circular rod with radius R_0, the longitudinal (U_1) and the transverse (U_3) displacements are coupled by

$$E_{33} = \nu U_{1,1} \tag{5.74}$$

that will cause additional stress

$$T_{11} = \frac{1}{2} \rho_0 \nu R_0^2 \ddot{U}_{1,1} \,, \tag{5.75}$$

where ν is the Poisson ratio. Using the standard approach for deriving the evolution equation (Nariboli and Sedov, 1970; Engelbrecht, 1983), we obtain

$$\frac{\partial u}{\partial \tau} + au\frac{\partial u}{\partial \xi} - \Omega^{-2} \frac{\partial^3 u}{\partial \xi^3} = 0 \,, \tag{5.76}$$

$$\Omega^2 = 3 \mid 1 + m_0 \mid \tau_c^2 u_0 l_0^{-2} c_0^{-1} \,, \tag{5.77}$$

where $l_0 = \nu R_0$ is the scale parameter and other notations given above. From that follows that dispersion due to the last term in the KdV equation is essential in rods of large radius for high-frequency inputs with small amplitudes. However, this model is too simplified.

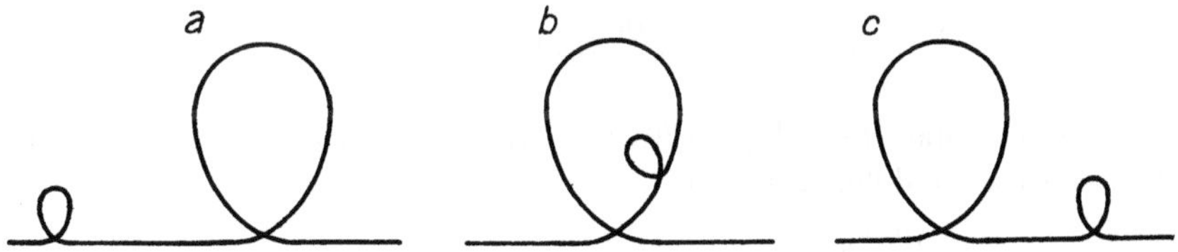

Fig. 5.1. Loop solitons for three successive time moments (from left to right).

The correct modelling brings us to a more complicated equation. Samsonov (1988, 1994) has shown that the longitudinal waves in a nonlinearly elastic circular rod are described by the equation

$$U_{tt} - c_0^2 U_{xx} = \frac{1}{2}[kU^2 + \nu^2 R_0^2 (U_{tt} - c_1^2 U_{xx})]_{xx} \tag{5.78}$$

where $U \equiv U_1, c_0$ is the usual wave velocity of longitudinal waves, $c_1^2 = \mu/\rho$, and k is the nonlinear coefficient. This so-called "double-dispersion equation" does not permit factorization into right and left propagating waves, i.e. into evolution equations. Nevertheless, Eq. (5.78) possesses the solitary-type solutions but only for a specific range of parameters. Equation (5.78) permits also the generalizations for the case of a rod with the variable cross-section. For details the reader is referred to Samsonov (1994) where also an experimental evidence for solitary waves in rods is given.

When a rod is very thin resisting only tension then it is called elastica. Wave propagation in strait and helix elasticas is an extremely interesting problem.

First, the transverse motion $w(X,t), X \equiv X_1$ in a straight half-finite $(X > 0)$ elastica (string) subject to an excitation of the end is derived by

$$\frac{\partial^2 w}{\partial t \partial X} + \operatorname{sign}\left(\frac{dX}{ds}\right) \frac{\partial^2}{\partial X^2} \left\{ \frac{\partial^2 w/\partial X^2}{[1 + (\partial w/\partial X)^2]^{3/2}} \right\} = 0 \tag{5.79}$$

where s is the arc length along the solution curve (Wadati et al., 1979). This equation has a solution in the form of the loop soliton propagating with an amplitude-dependent velocity (Fig. 1.3). Such a soliton has a remarkable property: the smaller the amplitude, the larger the velocity, i.e. the smaller loop overtakes the larger one (Fig. 5.1). On the contrary, the KdV solitons behave vice versa: a larger soliton overtakes the smaller one. Loop solitons are analysed in detail by Konno and Jeffrey (1994) and Jeffrey (1994).

Second, excitation propagating in an elastic helix (Fig. 5.2) may lead to interesting phenomena. Slepyan et al. (1995) have used the equation of motion

$$\frac{\partial}{\partial s}\left[F(s,t)\frac{\partial}{\partial s}\mathbf{R}(s,t)\right] = \rho_0 \frac{\partial^2}{\partial t^2}\mathbf{R}(s,t) \tag{5.80}$$

where $\mathbf{R}$ is the position vector, F – a non-negative tension force, s – the coordinate along the fibre. Again, the solitary waves are found, governed by Eq. (5.80). In

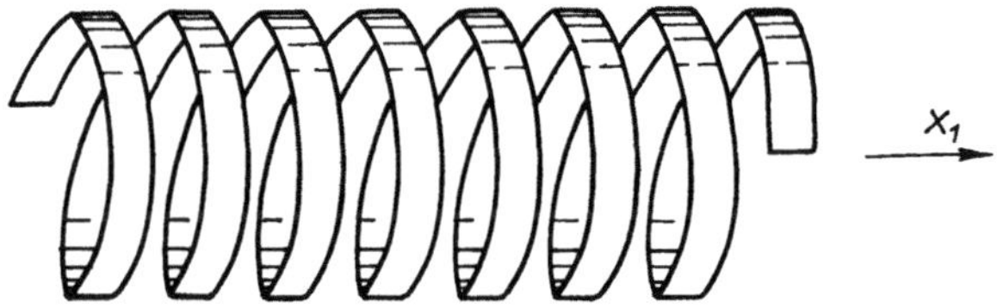

Fig. 5.2. A helix as a wave-bearing structure.

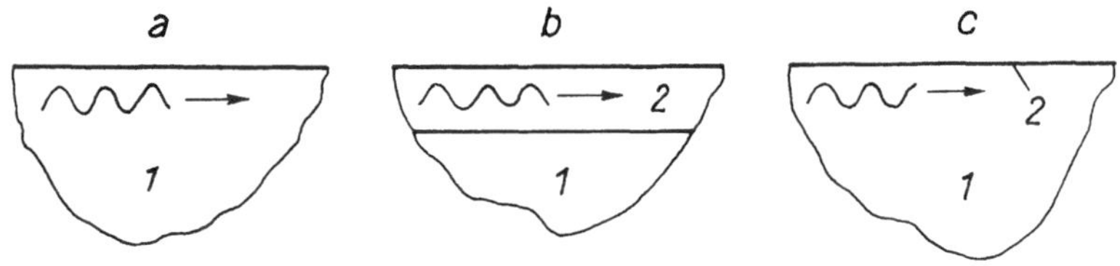

Fig. 5.3. Typical problems of surface waves: a – simple half-space 1; b – half-space 1 with a layer 2; c – half-space 1 with a thin film 2.

addition, complex static deformation patterns are known (Davies and Moon, 1993) involving quasiperiodicity and spatially stochastic deformation.

A very important class of wave phenomena occurs near a boundary. The typical object is a half-space where in addition to longitudinal and transverse waves, the surface waves occur. The energy of surface waves is spatially confined to the vicinity of the boundary. The problems under discussion are schematically shown in Fig. 5.3. The surface waves have been known for a long time but their nonlinear analysis has a much shorter history (see David and Parker, 1988). Recent results show that despite the complicated character of motion (the path of any particle in motion is an ellipse), there is a possibility to derive a nonlinear evolution equation for surface waves (Fig. 5.3a).

According to Parker (1994) one has

$$iJ\frac{\partial C(k,X)}{\partial X} + \int_{-\infty}^{\infty} \Lambda(k-\kappa,\kappa)(k-\kappa)C(k-\kappa,X)C(\kappa,X)\mathrm{d}k = 0\,. \qquad (5.81)$$

where $C(k,X)$ is the Fourier transform of the surface elevation and k is the wavenumber. The kernel $\Lambda(k-\kappa,\kappa)$ determines the contribution to the (complex) rate of change in $C(k,X)$ from each pair of wave numbers with sum k and involves also material nonlinearities. The coefficient J measures the strain energy associated with the displacement field. Similar results have also been obtained by Zabolotskaya (1992).

As mentioned in Chapter 3, nonlinearity in surface waves may be of nonlocal character depending on the penetration depth (Hamilton et al., 1995a). Introducing Hilbert transform

$$H[f] = \frac{1}{\pi} Pr \int_{-\infty}^{\infty} \frac{f(t')}{t'-t} \mathrm{d}t' , \tag{5.82}$$

where Pr denotes the principal value, the evolution equation for real horizontal displacement u of a nonlinear Rayleigh wave is (Hamilton et al., 1995b)

$$\frac{\partial u}{\partial \tau} - \tfrac{1}{8}\frac{\partial}{\partial \xi}(u^2) - \tfrac{1}{4} H\left[uH\left[\frac{\partial^2 u}{\partial \xi^2}\right]\right] = 0 . \tag{5.83}$$

Here, however, the "spatial" variable $\xi = t - X/c_R$ has the dimension of time while c_R denotes the velocity of the Rayleigh wave.

The distortion of wave profiles due to nonlinearity is essential in seismology as well as in acoustical devices based on surface waves.

Nonlinear waves in a layered half-space with one (Fig. 5.3b) or many layers are certainly more complicated to treat, but the general theory is known (Parker, 1994).

The case depicted in Fig. 5.3c is composed of a nonlinear isotropic elastic half-space and a superimposed linear elastic thin film (Maugin and Hadouaj, 1991; Maugin, 1994b). The thin film as an interface of zero thickness introduces dispersion into the system that together with nonlinearity of the half-space governs the surface waves. A typical governing system of equations in terms of the complex amplitude a of the SH mode and the real x-gradients $n_1 = v_x$, $n_2 = w_x$ of the longitudinal and transverse vertical components of the Rayleigh mode is

$$ia_t + a_{xx} \pm \lambda \mid a \mid^2 a + 2a(n_1 + n_2) = 0 , \tag{5.84}$$

$$(n_1)_{tt} - c_L^2 (n_1)_{xx} = -\mu_L (\mid a \mid^2)_{xx} , \tag{5.85}$$

$$(n_2)_{tt} - c_T^2 (n_2)_{xx} = -\mu_T (\mid a \mid^2)_{xx} . \tag{5.86}$$

Here λ is the coefficient of self-interaction; μ_L, μ_T are the mutual interaction coefficients; c_L and c_T are the characteristic velocities. This system is soliton-bearing, with a possible breather at the collision. For more details, the reader is referred to Maugin (1994b). The results are applicable in nondestructive evaluation techniques and signal processing by using acoustic devices.

One special case of surface-guided waves is the propagation along a wedge (Fig. 5.4a). The linear waves are nondispersive but nonlinear waves obey interactions between all the wave numbers (Parker, 1994). Despite the complicated geometry, even here one can derive the evolution equation. Antisymmetric modes, for example, are governed (Parker, 1994) by

$$\begin{aligned} iJ\frac{\partial C}{\partial Y}(k, Y) \;&=\; \int_{-x}^{x}\int_{-x}^{x} \Omega(k-\kappa-\nu, \kappa, \nu)(k-\kappa-\nu)\kappa\nu \times \\ &\times C(k-\kappa-\nu)C(\kappa)C(\nu) d\kappa d\nu = 0 , \end{aligned} \tag{5.87}$$

where $Y = \varepsilon^2 X_3$; $C(k, Y)$ as before is the Fourier transform of the wedge elevation; κ, ν are the wavenumbers and $\Omega(\sigma, \kappa, \nu)$ is the kernel. Parametric mixing and other effects are described by such an approach. This approach is also shown to be useful for analysis of waves in a half-space with the corrugated surface (Fig. 5.4b). This is

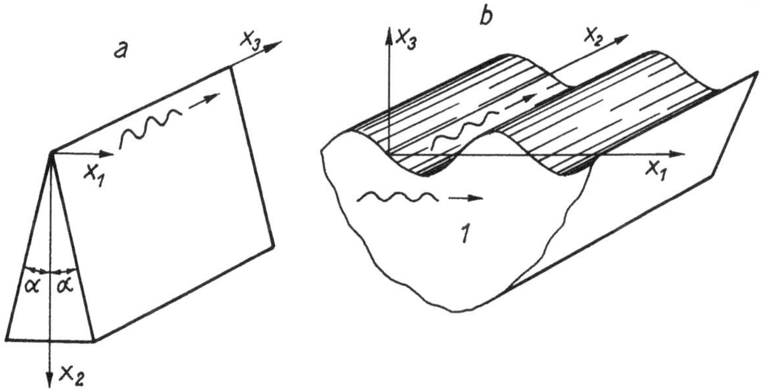

Fig. 5.4. Guided waves: a – a wedge; b – a half-space with corrugated surface.

an intriguing problem where much is to be done, but a 2D nonlinear Schrödinger equation is shown to serve as an evolution equation (see Parker, 1994)

$$i\frac{\partial a}{\partial y} + P\frac{\partial^2 a}{\partial z^2} + Q\frac{\partial^2 a}{\partial \tau^2} + R \mid a^2 \mid a = 0\,, \tag{5.88}$$

where $a(x_1, x_2, t)$ is the amplitude, y is the evolution coordinate, z is the transverse distance, and τ is the retarded time while, P, Q, R are related to the geometry.

5.2.2. Propagating instabilities

We return now to the definition of a wave as a state moving to another state. Defining instability as a state, it is easily concluded that a propagating instability is a wave. The question certainly is about the character of governing equations.

A good example to start with is the domino effect (Stronge, 1987). Consider a regularly spaced sequence of slender rectangular blocks (dominoes) standing on end with a small space λ between the elements. The dominoes stand vertically in the gravitational field. Toppling one element initiates a sequence of collisions (Fig. 5.5). If the initial energy is large enough, then a wave of collisions propagates over the entire array. This wave has a finite speed of propagation c determined from the condition of equality of the rates of diffusion and reaction (Stronge, 1989). One could say that "toppling is the reaction in a domino array that transforms a certain gravitational potential into kinetic energy" (Stronge, 1989). The governing equations are of the reaction-diffusion type that means quite another physical phenomenon (c.f. Maugin, 1990). Nevertheless, the outcome is a wave, propagating in a mechanical system with a finite speed. The speed can be calculated from the block spacing λ, the block width h and the period between collisions

$$c = (\lambda + \mu) / \int_0^{\psi} \dot{\theta}_1^{-1} d\theta\,, \tag{5.89}$$

where the angular velocity of the block $\dot{\theta}_1$ at the wavefront must be determined from the equation of motion. Reaction-diffusion type systems are more known in

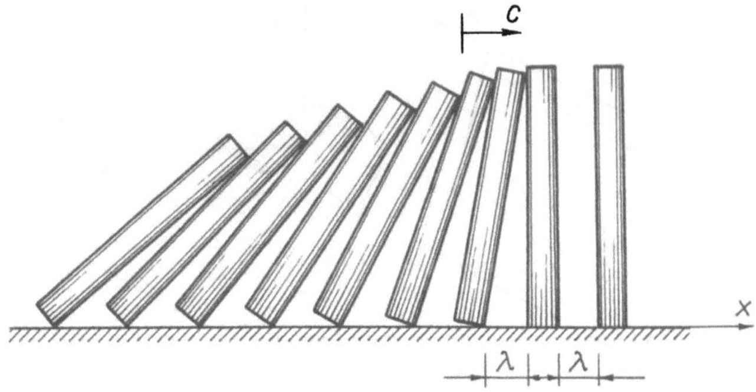

Fig. 5.5. Toppling dominoes; c, wave velocity.

chemistry and biology (Perelson et al., 1988), but this example shows a way to bridge such mechanical phenomena to other fields.

Noting that the toppling of one domino means generating an instability, we turn our attention to structures where instabilities may be buckles, bulges, discontinuities of deformation, or even of displacements, etc. An excellent review on propagating instabilities is presented by Kyriakides (1993). He treats the initiation and propagation of:

- bulges in inflated elastic tubes;
- buckles in long tubes and pipes under external pressure;
- buckles in long, confined cylindrical shells;
- and buckles in long shallow panels.

The common feature of structures exhibiting such instabilities is the local unstable response characteristic shown in Fig. 5.6. In certain sense the character of propagation is similar to that of toppling the dominoes. "Once the geometric integrity of such structures is compromised, the instability has the potential of spreading over the whole structure" (Kyriakides, 1993, p. 68). Such a situation is shown in Fig. 5.7. A pipeline is installed on the sea floor using a special vessel and if the technological characteristics of the process (movement of a vessel, loss of tension, etc.) are wrong, the pipe can buckle due to the combined effect of bending and pressure. Once initiated, the buckle can also propagate provided the external pressure is high enough. The governing equations for the beam-flexural mode A and the ring-flexural mode B (scaled over radius a) of the pipeline with a round cross section are (Sugimoto, 1989):

$$\rho_0 \frac{\partial^2 A}{\partial t^2} + \frac{Ea^2}{2} \frac{\partial^2}{\partial z^2} \left[\left(1 + \frac{3}{2} B + \frac{1}{16} B^2 \right) \frac{\partial^2 A}{\partial z^2} \right] = T \frac{\partial^2 A}{\partial z^2} , \tag{5.90}$$

$$\rho_0 \frac{\partial^2 B}{\partial t^2} + \frac{3Eh^2}{5a^4} \left(1 - \frac{p}{p_c} \right) B + \frac{Ea^2}{20} \frac{\partial^4 B}{\partial z^4} =$$

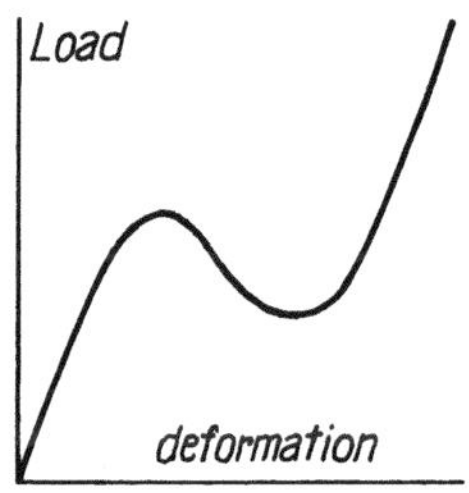

Fig. 5.6. Local unstable response.

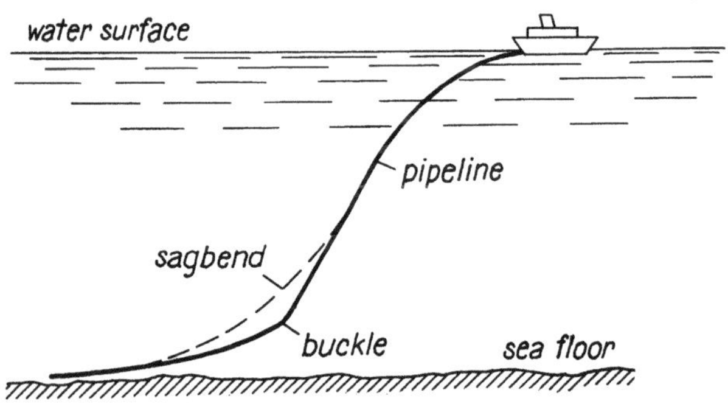

Fig. 5.7. Initiating of a buckle during an installation of an offshore pipeline.

$$= T_B \frac{\partial^2 B}{\partial z^2} - \frac{3E}{5}\left(1 + \frac{5B}{6}\right)\left(\frac{\partial^2 A^2}{\partial z^2}\right), \tag{5.91}$$

where z is the axial coordinate; ρ_0 and E are the density and Young's modulus, respectively; p and p_c denote the hydrostatic side pressure and the buckling pressure; h is the effective thickness of the pipe; and finally T_A and T_B denote the effective tension for both modes, respectively. The propagation velocity of a buckle is determined from the expression

$$\left(\frac{v}{v_0}\right)^2 = \left(\frac{h}{2a}\right)^2 \left[\frac{1}{2}(N + p/p_c) - \frac{2^{1/2}}{9S}(1 - p/p_c)^{1/2}\right], \tag{5.92}$$

where S and N denote the dimensionless maximum deflection and the scaled axial force, respectively, and $v_0^2 = E/\rho_0$. Note that the governing Eqs. (5.90) and (5.91) are of the dispersive character and nonlinearly coupled, and describe actually the ovalization of the cross section. A simple experiment showing initiation and propagation of a buckle can easily be performed with a flexible steel tape measure.

In general terms, the propagation of instabilities can be explained in terms of the global load-deformation curve (Fig. 5.8). Following Kyriakides (1993), the

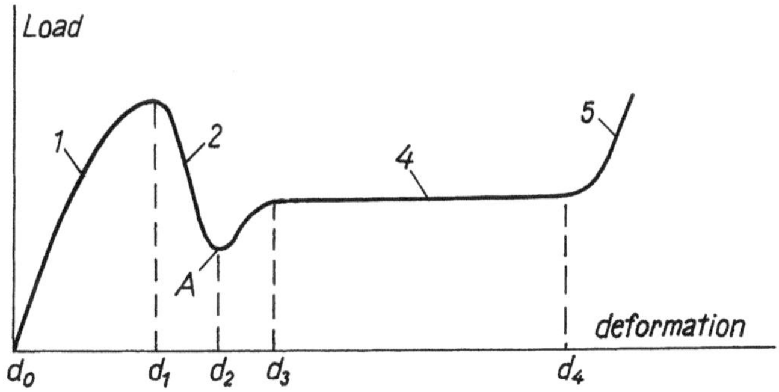

Fig. 5.8. Global load-deformation response: 1 – uniform deformation; 2 – localization; A – arrest; 4 – propagation; 5 – uniform deformation.

explanation follows. Initially, the structure deforms uniformly (region $d_0 - d_1$), that may mean also bifurcation at d_1. Then, in region $d_1 - d_2$, the instability is localized and at d_2 (a local minimum), it is arrested. Region $d_3 - d_4$ corresponds to the propagation. In this case, the load is lower than needed for the bifurcation at d_1.

With a certain flexibility we may also classify a phase boundary as an instability. Clifton (1993) reports on failure waves in glasses that are the propagating phase boundaries. There is experimental evidence that above a certain impact stress, there is a propagating boundary between low-spall-strength material close to the impact face and high-spall-strength material in the rest of the specimen. This phase change according to Clifton (1993) could be a transformation either to a crystalline phase or to an amorphous phase with a higher coordination number.

And last but not least, the propagation of the discontinuity of the displacement, i.e., crack. The propagation of a crack again corresponds to the definition of waves. Here, however, the plastic deformation at the crack tip should be taken into account. The finite velocity of the crack tip certainly depends on the stress state. For a straight expanding crack ($x < l(t)$), the velocity $\dot{l} < c_2$ where c_2 is the shear wave velocity, is determined by the following expression (Slepyan, 1990)

$$\dot{l}(t) = \left(c_2 - \frac{\pi K^2}{8\sigma^2 t}\right) H(t - t_*) , \tag{5.93}$$

$$t_* = \pi K^2 / 8\sigma^2 c_2 , \tag{5.94}$$

where K is the critical coefficient of intensity and σ is the uniform stress at the crack surface. Generally speaking, this velocity characterizes the propagation of the viscoplastic zone around the crack tip. Maugin (1994b) calls this structure a "viscoplastic soliton." The resemblance to the soliton motion in a conservative system is striking, including the amplitude dependence on the propagation velocity.

5.3. SOLITARY WAVES

5.3.1. Definitions

Solitary waves form a paradigm in many branches of physics, including wave propagation in solids. Discovered first in a "natural experiment" by John Scott Russel, the solitary waves were theoretically found later as steady state solutions to shallow water equations (Korteweg and de Vries, 1895). The quest for solitons actually started from the problem of energy equipartition in lattice dynamics—the Fermi-Pasta-Ulam problem—and the soliton formation from a harmonic input for the same problem was demonstrated at the continuum limit (Zabusky and Kruskal, 1965), restoring the Korteweg-de Vries (KdV) equation. The story about the soliton is presented in Section 7.2 with the selected list of references.

Solitary waves are usually defined on different levels of mathematical and physical accuracy. From the mathematical viewpoint, the solitary waves are excellent examples of solutions to conservative systems. From the physical viewpoint, they are sometimes referred to as "simple structures in complicated systems". This is the reason why after describing the models of wave propagation in complicated materials (Section 5.1) and structures (Section 5.2), we turn our attention to solitary waves treating them as a touchstone in the complexity of Nature.

Nature, however, is by far too rich in its complexity to be described only by one class of equations or systems, i.e. conservative systems including the KdV equation. Solitary waves may also occur in other situations which are not conservative, for example, in diffusive systems. Wave motion in complicated media may also be influenced by an additional energy influx caused by accompanying processes and the influence of coupled fields. This may lead to a r.h.s of the governing evolution equation. If the governing evolution equation without the r.h.s possesses solitary-type solutions, then the question of stability of these solutions must be analysed for the case of the non-zero r.h.s. We may also ask: is a solitary wave always a soliton, what is the difference between them if any, etc. This means to use a good ordered system of definitions.

We present here a rather detailed system of definitions based on that of Scott et al. (1973) and Engelbrecht (1990).

The hierarchical system is built up according to the principle—the more general first, the more specific later. Attention is paid to the fact that the formation of "simple structures in complicated systems" leads to stable profiles usually described by ordinary differential equations. In many cases, however, the exact type of the ordinary differential equation under consideration is not of primary importance when the properties of a solution are classified. The type of an equation is specified only when a specific property connected with a specific equation type is considered.

B a s i c n o t a t i o n s. As usual, x and t are space and time coordinates, and $x \equiv x_1$ is the space coordinate for one-dimensional problems considered here. The moving frame $\xi = x \pm ct, \tau = \varepsilon x$ (or $\tau = \varepsilon t$) is used for deriving the evolution equation (see Chapter 4). Here c =const, and ε is a small parameter used for stretching. Further another frame $\eta = \tau \pm V\xi, V$ =const. will also be used, and its physical meaning will be given.

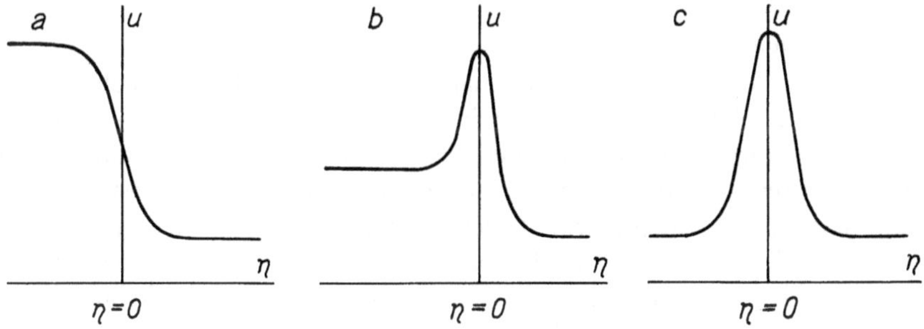

Fig. 5.9. Solitary waves with various equilibrium states.

D e f i n i t i o n 1: *Progressive steady waves are waves which depend on the space coordinate x and time t only through the variable* $\xi = x \pm ct, c =$const.

Note that in the case of multidimensional processes, generally speaking, $\xi \to \varphi(\mathrm{x}) \pm ct$. The sign of variable ξ has no principal meaning here and further $\xi = ct - x$, for example, is also used.

Def. 1 is widely used in linear theory where velocity c is either constant or depends on space coordinates. In nonlinear theory, however, its possible dependence on the solution (amplitude) should also be considered. In this case instead of Def. 1 one should use

D e f i n i t i o n 1 a: *Progressive steady waves are waves which depend on the coordinates* ξ *and* τ *only through the variable* $\eta = \tau \pm V\xi, V =$const.

The following hierarchy of definitions is based on Def. 1a.

The next step is to extract the localized structures. Assuming the condition of smoothness, let us introduce

D e f i n i t i o n 2: *Solitary waves are progressive steady waves localized at* $\eta = 0$*, the profiles of which describe the smooth transition from one constant limit state at* $\eta \to -\infty$ *to another constant limit state at* $\eta \to +\infty$.

The limit states need more explicit information. The most important case is given by

D e f i n i t i o n 3: *Solitary pulse waves are progressive steady waves localized at* $\eta = 0$*, the profiles of which describe a smooth transition from an equilibrium state at* $\eta \to -\infty$ *to the same equilibrium state at* $\eta \to +\infty$.

The examples of solitary waves are shown in Fig. 5.9. The profile in Fig. 5.9c corresponds to Def. 3.

A widely known solitary wave in contemporary mathematical physics is "soliton" (Zabusky and Kruskal, 1965). Actually, soliton has formed a new paradigm in physics, but the notion of soliton itself has been used in a rather broad sense. That is why there is a need for a "working" definition at least on the level of solitary waves. The most important property of a soliton is closely related to the interaction of solitons. This is emphasized by

D e f i n i t i o n 4: *Solitons are solitary waves which conserve their profiles and velocities during the collisions with other solitons.*

In many cases Def. 4 is sufficient in order to extract definite solitons from other solitary waves. The next step in the hierarchy is already determined by physical situations. There are several model situations with model evolution equations and explicit solutions, i.e. solitons. The main question is to establish the number of independent variables needed for describing a soliton. In physical terms this is the question how the propagation velocity is related to other parameters describing the process. Here we restrict ourselves to one-soliton solutions only, n-soliton solutions are discussed elsewhere (see, for example, Bullough and Caudrey, 1980).

The most celebrated and well-known evolution equation in nonlinear mathematical physics is the KdV equation derived for modelling one-dimensional dispersive waves in hydrodynamics, plasma physics, etc. (Korteweg and de Vries, 1895; Lamb, 1980; and many others). One of its possible presentations is

$$u_\tau + uu_\xi + u_{\xi\xi\xi} = 0 \,. \tag{5.95}$$

This equation has the widely known soliton-type solution

$$u = A\,\mathrm{sech}^2[(A/12)^{1/2}](\xi - A\tau/3 + \xi_0)\,, \tag{5.96}$$

where $A = \text{const.}$ is an amplitude and ξ_0 is the phase shift. Evidently there are two independent parameters A and ξ_0 in solution (5.96), and the propagation velocity is proportional to $A/3$. This is a solely nonlinear effect and leads to

D e f i n i t i o n 5: *Velocity of a KdV soliton depends upon the amplitude.*

In many cases, especially at the early stages of soliton dynamics, Defs. 4 and 5 have usually been put together and a soliton has always been treated as a KdV-soliton. Nowadays more specific definitions are needed and they will be given later in Section 5.3.4.

There is also another possible starting point related to the fact that many model cases describe the propagation in dispersive media with certain Lagrangians (Whitham, 1974). The existence of Lagrangians shows the importance of the energetical background. Starting from this viewpoint (Bishop et al., 1980) the following definition is justified:

D e f i n i t i o n 6: *Solitons are spatially compact structures with finite energy corresponding to a certain local energy minimum of a nonlinear field.*

Note that the identity properties at collisions (Def. 4) are not emphasized here, and tribute is paid to the "structure with finite energy". A typical example is a KdV soliton which (i) is a progressive steady wave (Def. 1), (ii) is a solitary pulse wave (Def. 3), (iii) is a "real" soliton in the sense of Defs. 4 and 6, and (iv) has certain special properties (Def. 5).

The next step is to combine the energetical viewpoint with the symmetry properties. The importance of being integrable in classical soliton theory (see, for example, Lamb, 1980) for KdV solitons is described in detail. A KdV soliton is symmetric with respect to $\eta = 0$ and the equilibrium state at $\eta \to \pm\infty$ is usually realized by $u \to 0$.

It is also quite logical to consider other structures which correspond to Def. 6 but are asymmetric. In this case $u \to +0$ at $\eta \to +\infty$ and $u \to -0$ at $\eta \to -\infty$, though a localized at $\eta = 0$ structure is preserved. Therefore one should distinguish two types of solitary pulse waves according to

D e f i n i t i o n 7: *Symmetric solitary pulse waves tend to the equilibrium state at* $\eta \to \pm\infty$ *either as* $u \to +0$ *or as* $u \to -0$.

D e f i n i t i o n 8: *Asymmetric solitary pulse waves tend to the equilibrium state at* $\eta \to +\infty$ *as* $u \to \pm 0$ *and at* $\eta \to -\infty$ *as* $u \to \mp 0$.

Generally speaking, asymmetric solitary waves are characteristic of nonconservative media, i.e. of media where an energy influx exists due to other processes (active media) (Engelbrecht, 1989). In this context it is useful to formalize the primary requirements:

S t a t e m e n t 1: *In closed systems energy is conserved through conservation law*

$$\mathcal{E}_t + P_x = 0\,, \tag{5.97}$$

where $\mathcal{E}$ is the energy density and P is the power flow (the indices denote the differentiation).

S t a t e m e n t 2: *In open systems the energy release is possible but the balance between the rate of energy release and other comparable effects may be stable.*

If the balance is stable (see Statement 2) then despite the fact that energy is not conserved, solitary waves may still exist in the system (see Section 6.4).

There is no need to enlarge the system of definitions beyond the basic ones given above. It may be useful, however, to notice two main classes of open systems: (i) with a weak energy influx (perturbed systems), and (ii) with a strong energy influx (Engelbrecht, 1991). In the case (i) the system is usually modelled by an evolution equation with a r.h.s. of a perturbative character, i.e. if the r.h.s. vanishes, the system will be conservative preserving solitary waves (Ostrovski, 1983). Here the following statement is useful:

S t a t e m e n t 3: *In open systems perturbed solitary waves may exist and preserve their properties at least for a bounded interval* $0 \le \tau \le \tau_{cr}$.

A perturbed KdV soliton may serve as an example (Ostrovski, 1983; Engelbrecht, 1991). In the case (ii) solitary waves exist only for the full system (see Section 6.4).

5.3.2. Derivation of the KdV equation

The KdV equation as a cornerstone of solitonics with its incredible richness is also the best example to demonstrate the various approaches used for its derivation. In other words, complexity of nature when viewed from a proper viewpoint, exhibits the same simpler pattern despite different languages used to describe complexity. Below we discuss several basic approaches for deriving the KdV as well as other "solitonic" equations.

Direct derivation. The classical KdV equation was derived by Korteweg and de Vries (1895) for waves in shallow water using directly the basic equations of fluid motion with physically admissible approximations. Their original equation had the following form

$$\frac{\partial \eta}{\partial t} = \frac{3}{2}\sqrt{\frac{g}{l}}\frac{\partial}{\partial x}\left(\frac{2}{3}\alpha\eta + \frac{1}{2}\eta^2 + \frac{1}{3}\sigma\frac{\partial^2 \eta}{\partial x^2}\right) \tag{5.98}$$

where η is the surface elevation, l is the depth of the canal, α and σ are the parameters depending on the capillary tension and density, respectively, and g is the acceleration due to gravity.

Discrete viewpoint. Contemporary understanding of solitary waves dates back to Zabusky and Kruskal (1965) who generalized the equations of lattice dynamics (finite number of freedom) to their continuum limit and obtained the KdV equation. The 1D lattice of length L was composed by $n-1$ identical masses m connected between each other by n nonlinear springs of length h. The equations of motion are then

$$my_{itt} = k(y_{i+1} - 2y_i + y_{i-1})(1 + \alpha(y_{i+1} - y_{i-1}))\,, \quad i = 1, 2, \ldots, n-1\,, \tag{5.99}$$

where $y_0 = y_n = 0$, y_i is the displacement of the i-th mass, k is the spring constant and α measures the strength of the nonlinearity of the spring. The force, exerted by compression or extension of the spring is

$$F = k(\delta + \alpha\delta^2)\,, \tag{5.100}$$

where δ is the relative displacement. Zabusky and Kruskal (1965) expanded y_i into Taylor series, set $kh^2/m = c^2$, $2\alpha h = \varepsilon$, $h^2/12\varepsilon = \mu^2$ and obtained

$$y_{tt} - c^2 y_{xx} = \varepsilon c^2 y_x y_{xx} + \varepsilon c^2 \mu^2 y_{xxxx}\,, \tag{5.101}$$

where terms of order higher than ε have been neglected. The next step was to use the simplest version of the asymptotic method (see Section 4.2) in order to arrive at the KdV equation in moving coordinates $\xi = x - ct$, $\tau = \varepsilon ct/2$ and properly chosen scaling of the amplitude q

$$q_t + 6qq_\xi + \mu^2 q_{\xi\xi\xi} = 0\,. \tag{5.102}$$

Nowadays the attention to discrete (lattice) model is again growing. The reason is obvious—materials used in contemporary high technology are very sensitive to changes in their microstructure that is best described by lattice models. An exhaustive review of such models is given by Maugin (1994b) keeping the pattern of Zabusky and Kruskal (1965). This means starting from a Hamiltonian and the equations of motion of discrete masses and then going to the continuum limit. Taking the simplest Born-Kármán model of a 1D lattice (chain) as a basic, Maugin (1994b) has many generalizations of this model related to various applications. The reader is referred to Maugin (1994b) for details, here we present just a short list of models:

(i) dynamics of lattices with transverse displacements resulting at continuum limit in equations of second-gradient elasticity (c.f. Section 5.1); these models are of importance in dynamics of shape-memory alloys and ferroelastic-martensitic transformations;

(ii) dynamics of defective lattices where the action of a defect may be considered as a substrate potential; the outcome is the Sine-Gordon or the Double Sine-Gordon equation;

(iii) dynamics of lattices with microstructure involving transverse displacements and rotations; this yields the models of the micropolar Cosserat medium (c.f. Section 5.1);

(iv) dynamics of lattices involving longitudinal and transverse displacements and two rotations; this is the case of elastic ferroelectrics and the continuum models involve coupled Sine-Gordon equations with possible simplifications.

Asymptotic methods. This is the well-grounded formalism based on the reductive-perturbative idea (Taniuti and Nishihara, 1983) applied to the mathematical models that are derived within the framework of continuum mechanics. The details can be found in Chapter 4.

Physical approach. This can actually be called the method of approximate (etalon) evolution equations and is based on the dispersion analysis (Whitham, 1974). Again the details may be found in Chapter 4.

Internal variable approach. This approach makes a clear distinction between solitonic and dissipative systems (Maugin, 1990). The general idea of this approach is given in Chapter 2.

Eshelby tensor formalism. This approach is developed in order to account for elastic inhomogeneities (Maugin, 1992a) that again cause the dispersion of waves (see Section 5.1).

The solitonic equations derived by various methods take rather complicated forms, such as

$$u_\tau + \beta(F(u))_\xi + \int_{-\infty}^{\infty} K(\xi - \eta)\frac{\partial u(\eta, t)}{\partial \eta} d\eta = 0\,, \tag{5.103}$$

$$u_\tau + \beta(F(u))_\xi + \gamma(uu_\xi)_\xi + b_3 u_{\xi\xi\xi} + b_5 u_{5\xi} + \ldots = 0\,, \tag{5.104}$$

where $F(u)$ is a nonlinear function, $\beta =$ const. and ξ, τ as a rule are the moving frame coordinates. The celebrated KdV equation is

$$u_\tau + \beta(F(u))_\xi + b_3 u_{\xi\xi\xi} = 0\,, \quad F(u) = \tfrac{1}{2}u^2\,. \tag{5.105}$$

The problem of integrability is quite important for Eqs. (5.103), (5.104). Further we demonstrate how the numerical simulation and the ideas of the Inverse Scattering Transform (IST) can be combined to get a detailed view on soliton formation.

5.3.3. Soliton formation in the KdV system

Statement of the problem We consider the standard KdV equation in the following form

$$u_t + uu_x + du_{xxx} = 0 \tag{5.106}$$

with the periodic boundary conditions

$$u(x,t) = u(x + 2n\pi, t)\,, \quad n = \pm 1, \pm 2, \ldots \tag{5.107}$$

For the sake of simplicity, throughout all this section the variables x and t are used instead of ξ and τ.

The initial excitation is given by

$$u(x,0) = -\sin x\,, \quad 0 \le x \le 2\pi\,. \tag{5.108}$$

The notation

$$d_l = -\log d\,, \tag{5.109}$$

where d is the dispersion parameter in Eq. (5.106), is introduced. The problem (5.106)–(5.108) is solved numerically for the following values of the parameter d_l:

$$d_l = 0, 0.1, 0.2, \ldots, 1.8, 1.85, \ldots, 2.25, 2.3209\,. \tag{5.110}$$

The value d_l corresponds to the value $\delta = 0.022$ used by Zabusky and Kruskal (1965). Note that $d = \pi^2\delta^2$. For numerical integration of Eq. (5.106) the pseudospectral method is used.

The idea and the essence of the Pseudospectral Method (PsM). The PsM was first proposed by Kreiss and Oliger (1972). Let the initial condition $u(x,0)$ be given on the interval 2π. The grid is formed by N points with $\Delta x = 2\pi/N$. The discrete Fourier transform (DFT) is defined by

$$U(\omega,t) = \mathrm{F}u = \sum_{j=0}^{N-1} u(j\Delta x, t)\exp(-2\pi i j\omega/N) \tag{5.111}$$

and the inverse discrete Fourier transform (IDFT) by

$$u(x,t) = \mathrm{F}^{-1}U = \sum_{\omega} U(\omega,t)\exp(2\pi i j\omega/N)\,, \tag{5.112}$$

where i is the imaginary unit and

$$\omega = 0, \pm 1, \pm 2, \ldots, \pm(N/2-1)\,,\ -N/2\,. \tag{5.113}$$

Here F denotes the Fourier transform and F^{-1} the inverse Fourier transform. Space derivatives are then given by

$$\frac{\partial u}{\partial x} = \mathrm{F}^{-1}(i\omega\mathrm{F}u)\,, \tag{5.114}$$

$$\frac{\partial^2 u}{\partial x^2} = -\mathrm{F}^{-1}(\omega^2\mathrm{F}u)\,, \tag{5.115}$$

$$\frac{\partial^3 u}{\partial x^3} = -\mathrm{F}^{-1}(i\omega^3\mathrm{F}u)\,, \tag{5.116}$$

$$\frac{\partial^n u}{\partial x^n} = \mathrm{F}^{-1}[(i\omega)^n\mathrm{F}u]\,. \tag{5.117}$$

In time, the finite difference leap-frog scheme is used. The KdV equation (5.106) leads to the following straightforward pseudospectral approximation

$$u(x,t+\Delta t) = u(x,t-\Delta t) - 2\Delta t u\mathrm{F}^{-1}(i\omega\mathrm{F}u) + 2d\Delta t\mathrm{F}^{-1}(i\omega^3\mathrm{F}u)\,, \tag{5.118}$$

with the linear stability condition

$$\frac{\Delta t}{\Delta x^3} < \frac{1}{\pi^3}\,. \tag{5.119}$$

Fornberg and Whitham (1978) proposed to replace $\Delta t\omega^3$ by $\sin\Delta t\omega^3$ in the last term of the expression (5.118), so the numerical scheme takes form

$$u(x,t+\Delta t) = u(x,t-\Delta t) - 2\Delta t u\mathrm{F}^{-1}(i\omega\mathrm{F}u) + 2d\mathrm{F}^{-1}[i\sin(\Delta t\omega^3)\mathrm{F}u]\,. \tag{5.120}$$

The linear stability condition for scheme (5.120) is

$$\frac{\Delta t}{\Delta x} < \frac{3}{2\pi^2}\,. \tag{5.121}$$

So the time step one can use for the scheme (5.120), proposed by Fornberg and Whitham (1978), is 4.71 times greater than for the classical scheme (5.118). Since $\sin\Delta t\omega^3 = \Delta t\omega^3 + O(\Delta t^3)$, the two methods are identical in the limit Δt decreasing to zero. The discussion about the accuracy of the schemes is given by Salupere (1995).

If we use (5.111) for calculating the DFT, then the spectral amplitudes are defined as

$$\begin{aligned} SA_k(t) &= \frac{2 \mid U(k,t) \mid}{N}, \quad k = 1,2,\ldots,N/2-1, \\ SA_k(t) &= \frac{2 \mid U(k,t) \mid}{N}, \quad k = N/2, \end{aligned} \tag{5.122}$$

and spectral densities as

$$\begin{aligned} S_k(t) &= SA_k^2, \quad k = 1,2,\ldots,N/2-1, \\ S_k(t) &= 2SA_k^2, \quad k = N/2. \end{aligned} \tag{5.123}$$

Next we introduce a cumulative spectrum as

$$SC_k = \sum_{i=k}^{N/2} S_i, \quad k = 1,\ldots,N/2. \tag{5.124}$$

Therefore one can interpret the quantities SC_k as summed energies (c.f. Bracewell, 1978).

Application of the IST for soliton formation. In the case of localized initial excitations, the problem to find emerging solitons is almost trivial (Kadomtsev, 1988). However, for periodical initial excitations, the problem is much more complicated. The analytical method of solving the KdV equation, the IST (Gardner et al., 1967) is rather sophisticated. The papers devoted to the study of the periodical initial conditions (e.g. Abe and Abe, 1979; Salupere et al., 1994) are mostly numerical or semi-numerical. Here we present in detail a very simple approach to this problem, based on the basic ideas of the IST method (Kalda, 1995).

As earlier, we present the KdV equation in the form

$$u_t + uu_x + du_{xxx} = 0 \tag{5.125}$$

and let the initial condition of the function $u(x,t)$ be given by

$$u(x,0) = \cos(x). \tag{5.126}$$

According to the IST (Gardner et al., 1967), the eigenvalues λ of the associated Schrödinger equation

$$-6d\psi_{xx} - u(x)\psi = -\lambda\psi \tag{5.127}$$

are constant in time. Besides, each isolated eigenvalue is associated with a soliton of amplitude 2λ. The number of solitons can be found as the number m of isolated eigenvalues $\lambda_0, \lambda_1, \ldots, \lambda_{m-1}$ of the associated Schrödinger equation. In the case of localized initial profiles of the function $u(x,t)$, this procedure is rather straightforward. Unfortunately, in the case of periodic initial conditions the isolated eigenvalues are absent. Our suggestion is to consider a large but finite number of periods. Thus let the initial condition be

$$u(x,0) = \cos(x)\Theta(A-x)\Theta(A+x), \tag{5.128}$$

where $A \gg 2\pi$ and $\Theta(x)$ is the Heaviside function.

The eigenfunctions are defined as the bounded solutions of Eq. (5.128), $\psi(x)$, $x \in R$. Due to the symmetry with respect to the transformation $x \to -x$, the eigenfunctions are either even or odd and can be found as the bounded solutions of the Cauchy problem

$$6d\psi_{xx} + [\cos(x)\Theta(A-x)\Theta(A+x) - \lambda]\psi = 0\,, \tag{5.129}$$

$$\psi(0) = 1\,, \quad \psi_x(0) = 0\,, \tag{5.130}$$

or that of Eq. (5.129) and

$$\psi(0) = 0\,, \quad \psi_x(0) = 1\,. \tag{5.131}$$

It is easy to see that the largest eigenvalue is less than one. Indeed, in the case of $\lambda = 1$ the solution $\psi(x)$ of the problem (5.129), (5.130) is exponentially diverging and with no zeros since the ratio ψ_{xx}/ψ is always positive. According to the general properties of the linear differential equations of the second order if we have two functions $\psi^a(x)$ and $\psi^b(x)$ such that within a certain interval of x, $\psi^a_{xx}/\psi^a \le \psi^b_{xx}/\psi^b$, then the function $\psi^a(x)$ has at least as many zeros as the function $\psi^b(x)$.

Evidently, decreasing the parameter λ leads to the appearance of the first, second, etc. zeros of the function $\psi(x)$. Let the transition point between the cases of one zero and without zeros be at $\lambda = \lambda_0$. The respective solution of the system (5.129), (5.130) does not have zero points and vanishes at infinity, i.e. it is the eigenfunction of the largest eigenvalue λ_0. A qualitative plot of it is depicted in Fig. 5.10a. The other eigenvalues can be found in a similar manner. It is convenient to enumerate the eigenvalues in the decreasing order of the values, starting to count from zero. Then the order number of an eigenvalue is equal to the number of zeros of the respective eigenfunction.

By small values of x, $x \ll A$, the function is almost 2π-periodic, with slowly decreasing amplitude of the periods (see Fig. 5.10). So we can conclude that

$$\psi^0_x|_{x=\pi} \approx \psi^0_x|_{x=-\pi} = -\psi^0_x|_{x=\pi}\,, \tag{5.132}$$

i.e. the value $\psi^0_x\big|_{x=\pi}$ is very close to zero (though negative). Now let us consider the eigenfunction of two zeros $\psi^2(x)$. At small values of x it is very similar to the function $\psi^0(x)$. Somewhere at $0 < x < A$ its sign changes and further the amplitudes of the oscillations start increasing. Later, by $x \approx A$ the amplitudes again decrease. So we have $\psi^2_x|_{x=\pi} \approx \psi^0_x\big|_{x=\pi} \approx 0$ and hence $\lambda_2 - \lambda_0 \ll \lambda_0$. At the limit $A \to \infty$ we have $\lambda_2 \to \lambda_0$.

Now let us skip $2N-2$ eigenvalues, $N = A/2\pi$ being the number of periods. The eigenvalue λ_{2N} is associated with the $2N$-th soliton, thus we have skipped two solitons per period. The respective eigenfunction ψ^{2N} has $2N$ zeros: it changes sign twice during each period and is almost periodic by small values of x. Analogously to the case of ψ^0, we have again $\psi^{2N}_x\Big|_{x=\pi} \approx 0$. The order number of the respective soliton is also $2N$; dividing it by the number of periods we infer that there are two solitons per period, larger than this one. So, it is the third largest soliton of a period.

Naturally, there are many eigenfunctions ψ_n, which do not satisfy the condition $|\,\psi^n_x|_{x=\pi}\,| \ll 1$. These eigenfunctions correspond to the solitons which have origin close to the boundary of the initial signal ($x \approx \pm A$) and are due to the boundary effects. At the limit $A \to \infty$ we can neglect the boundary effects and thus the

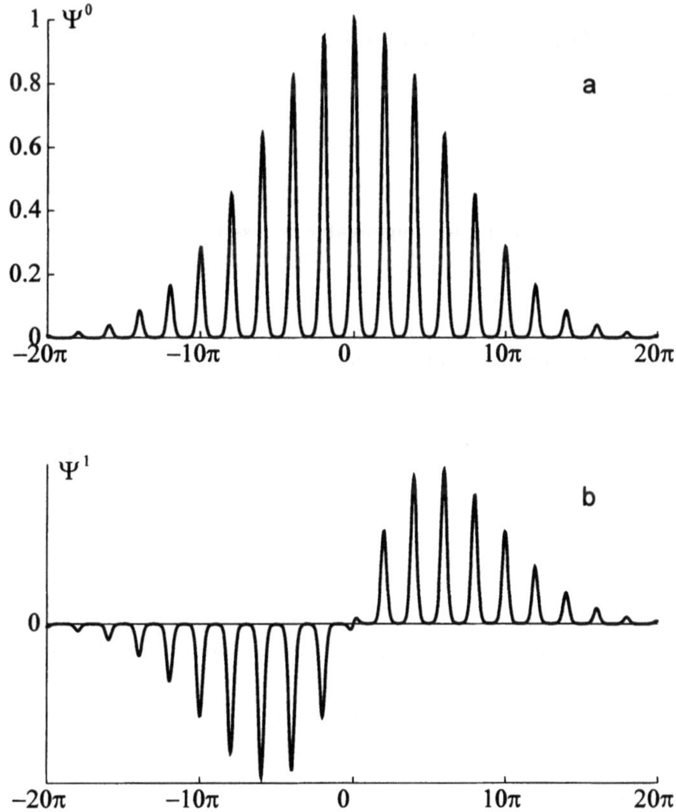

Fig. 5.10. Qualitative plot of the zeroth (a) and first (b) eigenfunctions of Eqs. (5.127), (5.128). The number of periods of the initial excitations is $N = 20$.

eigenfunctions of our interest are the 0th, $2N$th, etc. At the limit $A \to \infty$ they satisfy the following condition

$$\psi(0) = 1\,, \quad \psi_x(0) = 0\,, \quad \psi_x(\pi) = 0\,. \tag{5.133}$$

The procedure described above can be repeated in a similar manner for solitons of odd order number (a qualitative plot of the function ψ^1 is presented in Fig. 5.10b). The respective conditions can be written as

$$\psi(0) = 0\,, \quad \psi_x(0) = 1\,, \quad \psi_x(\pi) = 0\,. \tag{5.134}$$

Thus, if we enumerate the solitons originating from a single period of the initial excitation in the decreasing order of their amplitudes, the order number n of a soliton will be given by the number of zeros of the solution of the system (5.127),(5.132) or (5.127), (5.134) in the interval $[-\pi, \pi]$. The amplitude of the soliton can be calculated as $h_n = 2(\lambda_n - u_0)$. Here u_0 designates the reference level and will be defined later. In Fig. 5.11 we have depicted the first twelve eigenfunctions in the case of $\alpha = 10^{-2.3209}$.

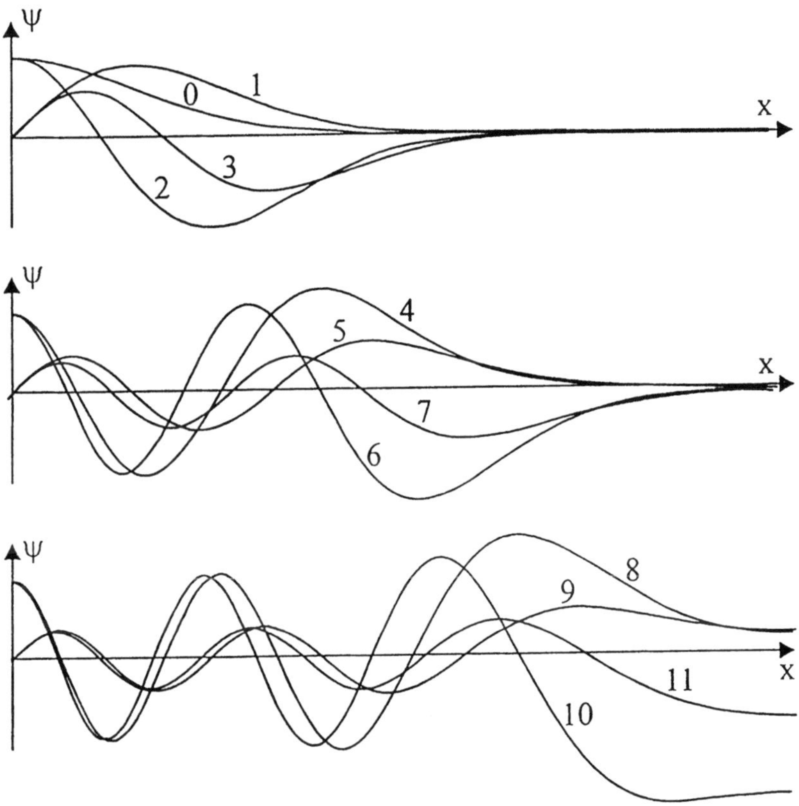

Fig. 5.11. The eigenfunctions of the associated Schrödinger equation.

That value has been used in the pioneering paper by Zabusky and Kruskal (1965). These curves were obtained via numerical integration of Eqs. (5.127)–(5.132) and (5.127), (5.134) by the Runge-Kutta method. The respective eigenvalues (the results of the same computation) are presented in Table 5.1.

In the case of a large but finite signal length A the negative values of λ are not allowed since the latter belong to the continuous spectrum of the associated Schrödinger equation. Thus there would be no solitons corresponding to the negative eigenvalues at the long time limit, such kind of solitons would decay at the time scale $\tau = A$. However, by strictly periodic initial conditions these solitons can be observable. They only disappear when the ambient level of the signal becomes higher than their eigenvalue, but later they become visible, again. In a certain sense these are virtual solitons. This conclusion can be confirmed by numerous computations, see e.g. Zabusky and Kruskal (1965).

A well-defined reference level u_0 is needed because then we could declare that the n-th soliton of eigenvalue λ_n is not observable if $\lambda_n < u_0$. There is also another reason

TABLE 5.1. The eigenvalues of the associated Schrödinger equation by $\alpha = 10^{-2.3209}$.

n	λ_n	n	λ_n	n	λ_n	n	λ_n
0	0.882115	1	0.650104	2	0.425819	3	0.209716
4	0.002353	5	-0.195561	6	-0.383082	7	-0.558783
8	-0.719759	9	-0.858305	10	-0.964385	11	-1.055964

why such a level is looked for. Namely, the height of a soliton is given by its eigenvalue λ_n: $h_n = 2\lambda_n$. Introducing the reference level corresponds to the transformation $u \to u - u_{\text{ref}}$. Thus the value of the reference level u_{ref} affects the analytical values of the heights of the solitons (both the absolute values and the values calculated with respect to the reference level u_{ref}). However, the different definitions of the reference level can be more or less equally founded and the particular choice depends on the character of the phenomenon to be studied. For instance, one can take u_{ref} equal to the minimal value of $u(x)$, averaged over time. The numerical data would give rise to $u_{\text{ref}} \approx -0.67$. The latter value has been used for instance by Salupere et al., (1994). It corresponds rather well to the first visual impression we get when we look at the plot of the function $u(x,t)$. Indeed, in the case of $\alpha = 10^{-2.3209}$, there are seven solitons of eigenvalue larger than -0.67 (Table 5.1); it is easy to identify all the seven solitons in the plot of the function $u(x,t)$ at any moment of time t.

A possible alternative could be to take $u_{\text{ref}} = -1$, the lowest possible value of the function $u(x,t)$. In the above mentioned case of $\alpha = 10^{-2.3209}$, such a reference level would correspond to eleven solitons (Table 5.1). The numerical results (c.f. Salupere et al., 1994) indicate that the solitons of eigenvalues between -1 and -0.67 are also visible, however some of them can be noticed only during short periods of time.

From the physical point of view it could seem that taking the reference level below the value $u = -1$ is senseless since the function $u(x,t)$ never descends down to such a value. However, according to Salupere et al. (1996), the use of spectral methods makes it possible to register even the solitons of eigenvalues slightly less than -1. A closer study of the behaviour of these "imaginary" solitons would be an intriguing problem.

Numerical analysis. In this section we apply PsM to integrate Eq. (5.106). The following notation is introduced: t_0 is the initial moment; t_{10} corresponds to the time when the initial spectral state is almost restored (i.e. $S_1 \to \max(S_1 \sim 1), S_i \to \min(S_i \sim 0)$ for $i > 1$) and $t_2 - A_0 \max[SA_2(t)]$ for $t_0 < t < t_{10}$; t_{ij} is the time when jth maximum of the i-th spectral amplitude is reached. It is clear that t_{10} corresponds to the recurrence time t_R defined by Zabusky and Kruskal (1965). Our final calculation time is denoted by t_f which is $1.6t_{10}$ for most values of d_l. It is

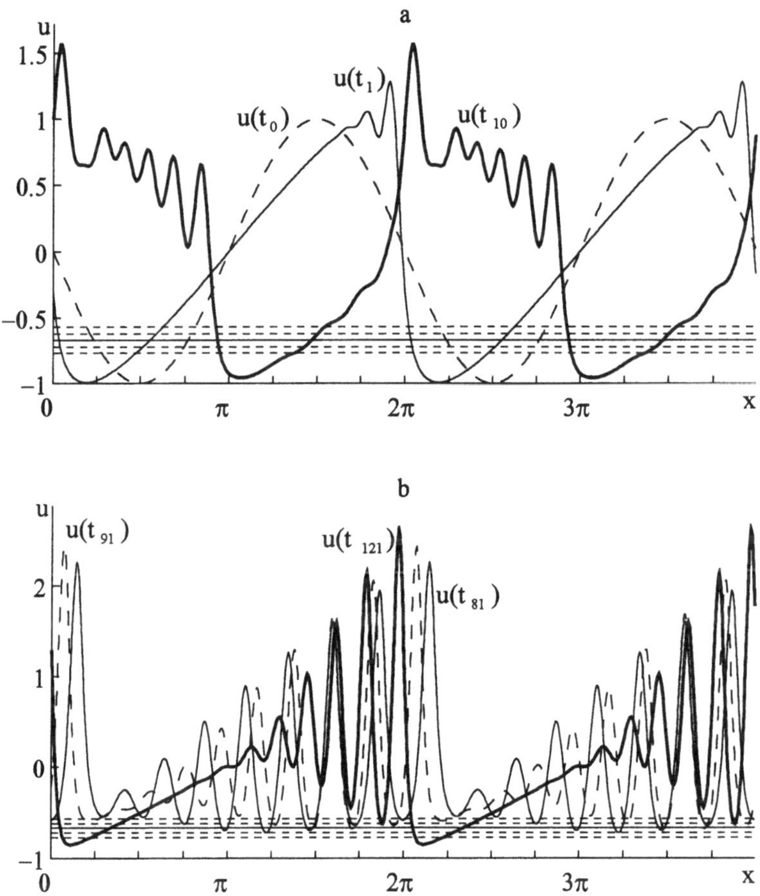

Fig. 5.12. Wave profiles at certain time moments t (see the text).

also clear that $t = 1$ is the breakdown time, i.e. the time when the initial wave $u(x,0) = \sin x$ breaks down for the nonlinear nondispersive ($\mu = 0$) equation.

The direct computation gives the following results. Here we present the full set of data for one dispersion parameter $d_l = 2.3209$ only with $t_f = 1.17t_{10}$. Results for other dispersion parameters d_l's are presented by Salupere et al. (1994, 1996).

The wave profiles at different t are depicted in Fig. 5.12. Here and further, the profiles in space are shown over two periods. The zero level $u_0 = -0.67$ is shown by a solid line with values $u_0 \pm 0.05, u_0 \pm 0.1$ shown by dotted lines. A time slice plot is certainly more informative. This is shown in Fig. 5.13. Bold lines indicate the profiles corresponding to certain t_i's or t_{ij}'s. Figure 5.14 presents the contour plot in the $x - t$ plane. Trajectories of solitons are marked by dots, a vertical dashed line marks the bound between two periods and horizontal dashed lines indicate the time

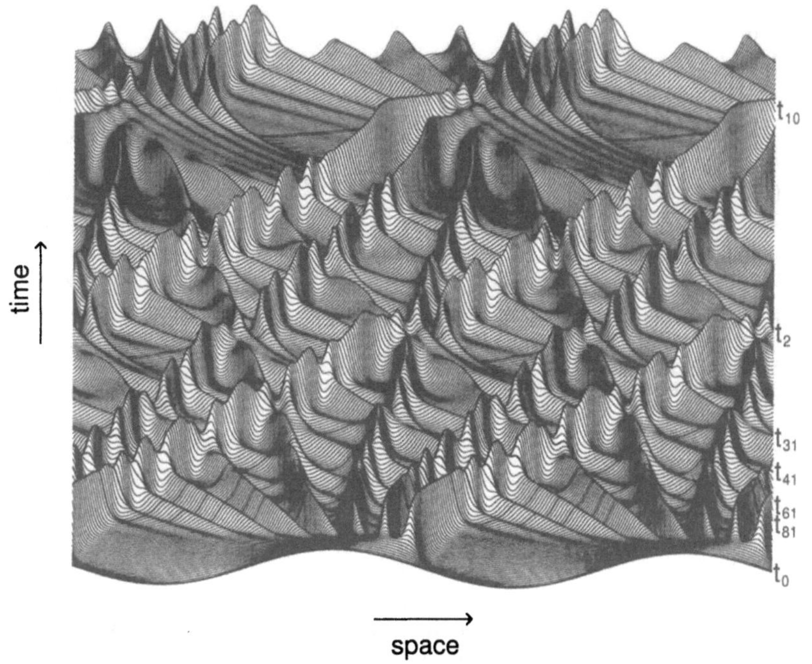

Fig. 5.13. Time slice plot.

moments t_i and t_{ij}. Solid contour lines are plotted with the step

$$\Delta u = (\max_{t,x} u - u_0)/10 . \qquad (5.135)$$

In Figs. 5.13 and 5.14 the different velocities of single solitons, their forward and backward phase-shifts and amplitude changes are clearly visible.

The time dependence of soliton amplitudes $u_a = u - u_0$ is shown in Fig. 5.15. The amplitude curves are also seen in Fig. 5.13, but here every soliton is traced with its own velocity. Every interaction is marked with a local minimum and these curves may also be called the resonance curves. Not all the solitons are visible all the time, that is often the case of solitons with smaller amplitudes.

Time plots of spectral densities $S_k, k = 1, 2, \ldots, 12$ are depicted in Fig. 5.16, while the cumulative spectrum is shown in Fig. 5.17. The constant value $SC1$ means that the second conservation law is satisfied.

For the forthcoming analysis the following properties of above results must be taken into account:

- the number of solitons;
- amplitude (resonance) curves and their maxima and minima;
- existence of local minima and maxima in time dependencies of spectral densities (spectral curves);
- convexity and concavity of spectral curves;
- relative and absolute values of spectral densities;

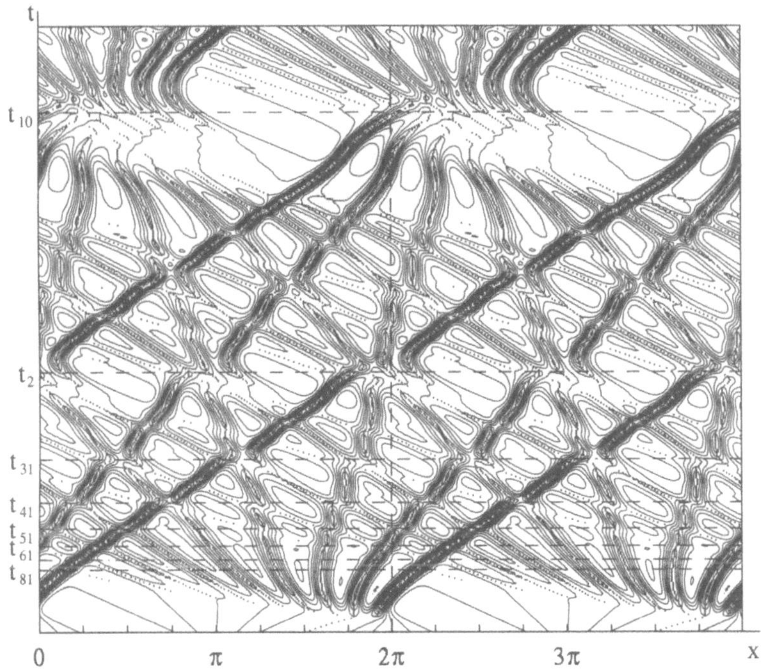

Fig. 5.14. Contour plot.

– relative and absolute values of minima and maxima of the cumulative spectrum.

The analysis presented below is based on studies of these properties. We hope to answer essentially two questions:

– how does the number of emerging solitons depend on the dispersion parameter and spectral characteristics?

– how is the soliton formation time reflected in the dynamics of spectral characteristics?

Number of emerging solitons. The number of emerging solitons from an arbitrary initial condition has the great importance for understanding the energy sharing between the modes. From the viewpoint of the IST, the number of emerging solitons is related to the eigenvalues of the corresponding Schrödinger equation, i.e. to the bound-state energy (see above). For bounded inputs the estimates are sufficiently accurate. For example, an initial excitation in the form of a rectangular well of width l and depth A results in N solitons, where

$$N = \text{largest integer } \leq S/\pi + 1\,, \quad S = lA^{1/2} \tag{5.136}$$

and for a sech^2 pulse of amplitude A and width l we obtain

$$N = \text{largest integer } \leq \tfrac{1}{2}\left[(1 + 4S^2/\pi^2)^{1/2} + 1\right]\,, \quad S = \pi l A^{1/2}\,, \tag{5.137}$$

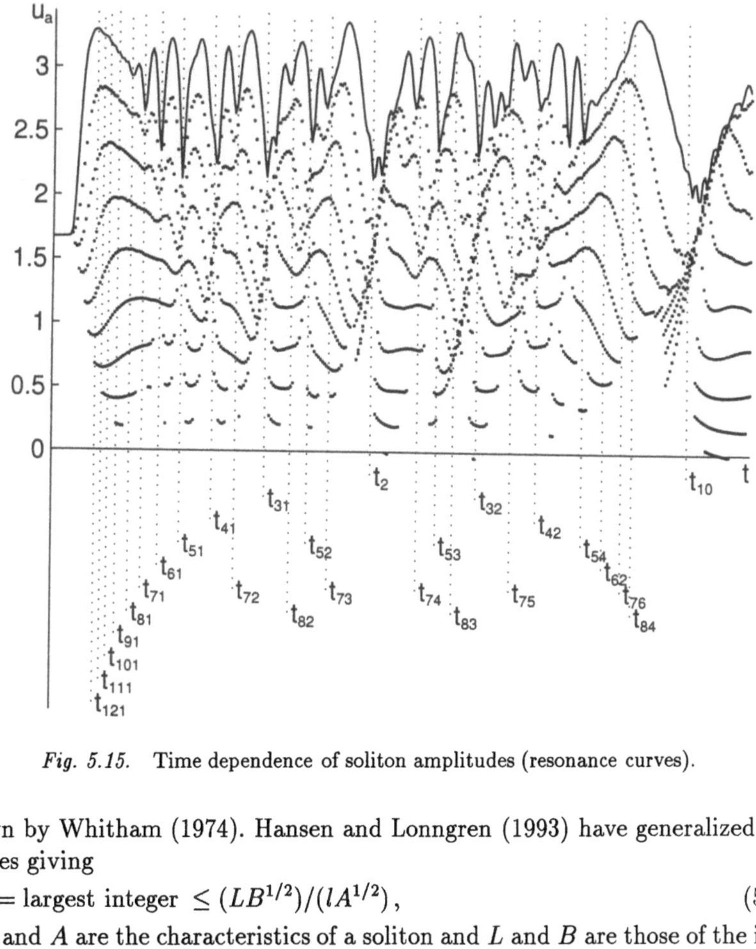

Fig. 5.15. Time dependence of soliton amplitudes (resonance curves).

as shown by Whitham (1974). Hansen and Lonngren (1993) have generalized these estimates giving

$$N = \text{largest integer } \leq (LB^{1/2})/(lA^{1/2}), \tag{5.138}$$

where l and A are the characteristics of a soliton and L and B are those of the input. This result is based on intuitive thinking rather than on theoretical basis. Abe and Satofuka (1981) have established the approximate number for solitons arising from a harmonic input $a \cos \pi x$. Given the dispersion parameter μ, the number of solitons is

$$N = 0.88U^{1/2}, \quad U = a/2\mu. \tag{5.139}$$

This formula is actually derived from the IST for a single-pulse initial excitation but is also applicable for the harmonic input (Abe and Satofuka, 1981). Zabusky's (1981) estimate

$$N \sim 0.2\delta^{-1} \tag{5.140}$$

coincides with Eq. (1.139).

As shown by numerical calculations, the zero-level for an initially monochromatic wave is not fixed in time and may for a certain time interval be below -0.67 that characterizes the fully formed train of solitons. Then the potential is deeper and a new soliton may emerge. We may call these solitons virtual. In the numerical

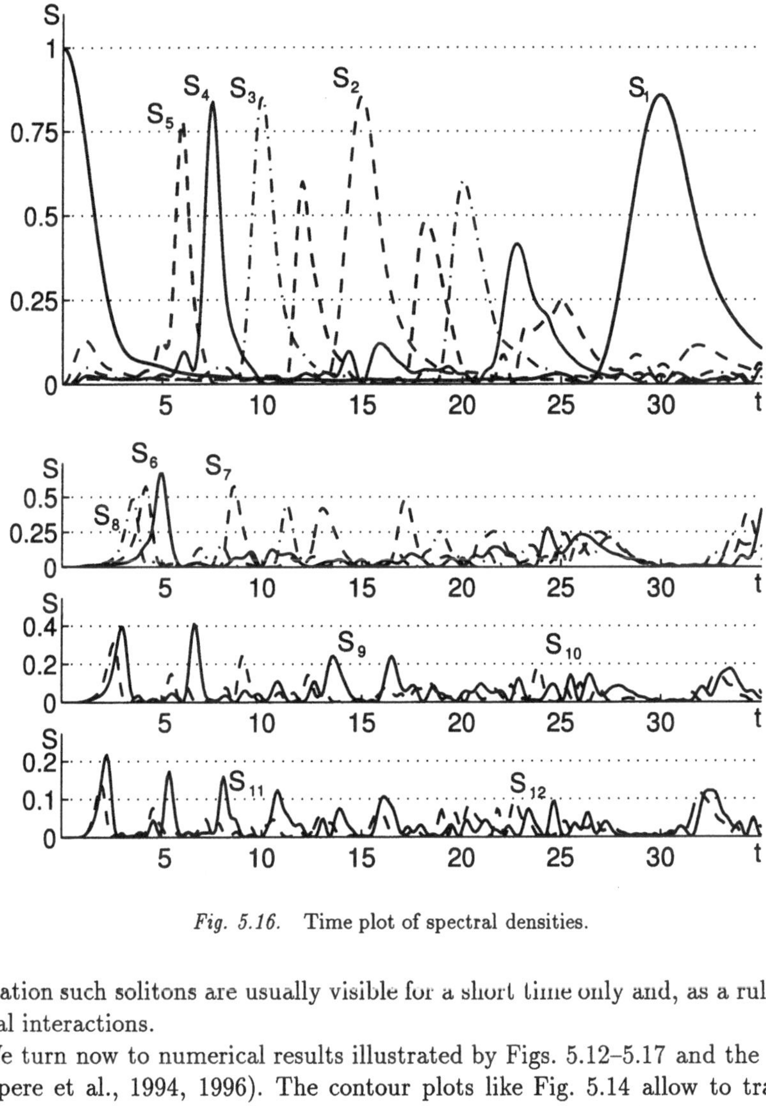

Fig. 5.16. Time plot of spectral densities.

simulation such solitons are usually visible for a short time only and, as a rule, after several interactions.

We turn now to numerical results illustrated by Figs. 5.12–5.17 and the full set (Salupere et al., 1994, 1996). The contour plots like Fig. 5.14 allow to trace the trajectories of the solitons, while the resonance curves (Fig. 5.15) show the solitons not only by different curves but also by interactions. It can be established that the number of interactions between solitons j and k ($j < k$) during the time interval $0 \leq t \leq t_{10}$ is equal to $k - j$. If n solitons form then the total number of interactions for the 1st (largest) soliton is

$$NI_1 = \tfrac{1}{2}(n-1)n \tag{5.141}$$

and for the mth soliton

$$NI_m = \tfrac{1}{2}(n-m)(n-m+1)\,. \tag{5.142}$$

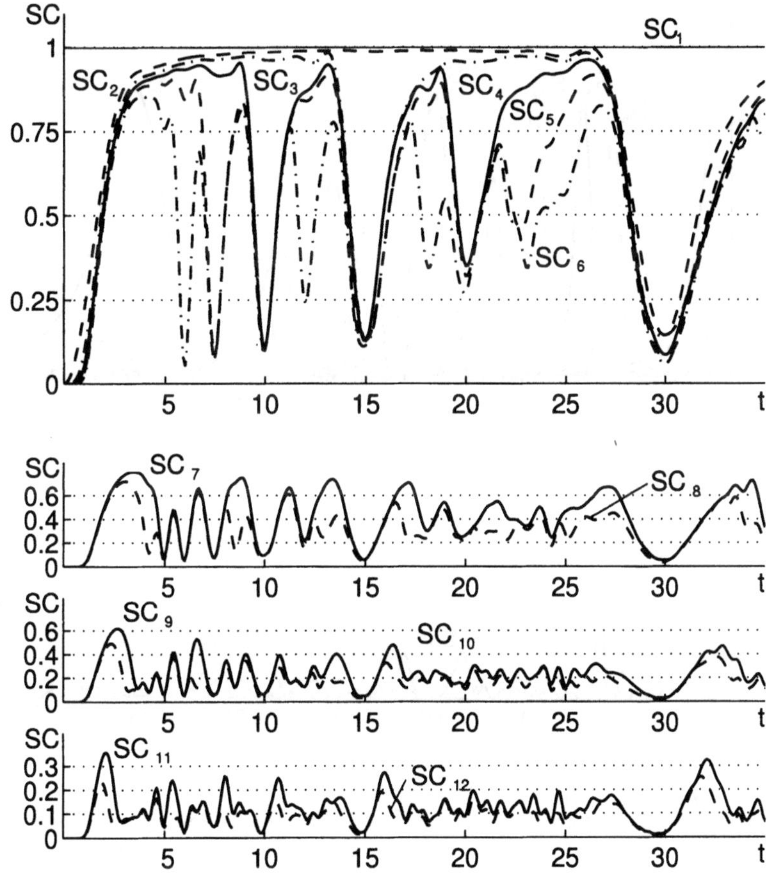

Fig. 5.17. Time plot of cumulative spectrum.

The total number of interactions between all n solitons during the time interval $0 \leq t \leq t_{10}$ is

$$NI_T = \tfrac{1}{6}(n-1)n(n+1)\,. \tag{5.143}$$

The results of this analysis are all summed up in Table 5.2. These data demonstrate clearly certain discrepancies but show also definite tendencies. First, formula (5.139) gives a rather good estimate, sometimes below, sometimes above the real number. Second, the IST certainly results in a theoretically correct number, but for an initially monochromatic wave the problem is to estimate the zero-level. For a train of solitons with -0.67 the exact number is 8 (c.f. Zabusky and Kruskal, 1965) but later the initial wave profile is almost restored and then we have to count with $u_0 = -1.0$. In this case, more solitons can be formed (see Table 5.2) and this is also continued by the analysis of amplitude curves and contour plots. The latter data also

TABLE 5.2. Number of solitons

d_l	Formula (5.139)	Contour plot		Amplitude curves	IST		Virtual solitons
		$t_0 < t < t_2$	$t_0 < t < t_j$		$u_0 = -0.67$	$u_0 = -1.0$	
0.00	1	1	1	2**	1	1	
0.10	1	1	1	2**	1	1	
0.20	1	1	1	2**	1	1	
0.30	1	1	1	2**	1	1	
0.40	1	2	2	2*	1	1	
0.50	1	2	2	2	1	2	1
0.60	1	2	2	2	1	2	1
0.70	1	2	2	2	1	2	1
0.80	2	2	2	3**	2	2	
0.90	2	2	2	3*	2	2	
1.00	2	3	3	3	2	3	1
1.10	2	3	3	3	2	3	1
1.20	2	3	3	3	3	3	
1.30	3	3	3	4	3	4	1
1.40	3	4	4	4	3	4	1
1.50	3	4	4	5*	3	4	1
1.60	4	4	5	5	4	5	1
1.70	4	5	5	6*	4	5	1
1.80	5	5	6	6	5	6	1
1.85	5	5	6	7*	5	6	1
1.90	6	6	6	7	5	7	2
1.95	6	6	7	8*	5	7	2
2.00	6	6	7	8	6	8	2
2.05	7	7	7	9*	6	8	2
2.10	7	7	8	9	6	9	3
2.15	7	7	8	10*	7	9	2
2.20	8	8	9	10	7	10	3
2.25	8	8	9	11*	8	10	2
2.32	9	9	10	12*	8	11	3

* *zero-level slightly (or order $O(\varepsilon)$) below* -1.0;
** *number of minima while velocities show a zero-level below* -1.0.

include the virtual solitons. These small solitons are seen for a small time interval only when the larger solitons due to their phase shifts give enough space. We call these solitons virtual. Such a situation seems to be characteristic of periodical initial conditions. The larger d_l, i.e. the smaller dispersion parameter μ, the more virtual solitons exist. Actually Zabusky and Kruskal (1965) exhibit their famous train of 8 solitons while the trajectories show 9 solitons already. Abe and Abe (1979), however, demonstrate the trajectories of 10 solitons for the same dispersion parameter.

The next question is: how is the number of solitons reflected in spectral characteristics? The plots of spectral densities (Fig. 5.15) show that some of the peaks are suppressed (see also Abe and Satofuka, 1981). A detailed inspection indicated that the lower harmonics suppress the higher ones if these are generated. A naive explanation may be given in terms of multiple unperturbed modes (1/2 vrs 2/4, 3/6, 4/8,...; 1/3 vrs 2/6, 3/9,..., etc.) that is depicted in Fig. 5.18 as a harmonics tree. The sequence of unperturbed modes can be represented as the Farey tree (Schroeder, 1991). However, due to interaction of solitons, the higher harmonics may still be visible later when the harmonics have been shifted. Two conclusions follow here. First, every first peak is of importance and, second, the harmonics S_j where j is a prime number, suffer less from interactions at least at the beginning of the process.

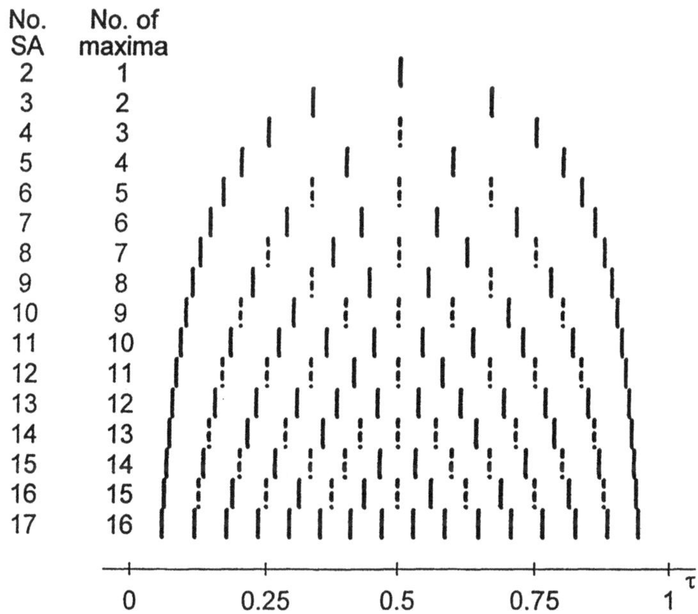

Fig. 5.18. Harmonics tree. Bold lines mark the maxima of harmonics without interactions; dotted lines – the supressed maxima; τ – normalized time.

The behaviour of spectral amplitudes and spectral densities in time, dependent on the dispersion parameter, is shown in Fig. 5.19. As mentioned already before, the concavity of SA_i curves was observed. By using a standard approach for the detection of extreme of a function, it is established that the concavity in SA_{k-1} appears in the neighbourhood of $t = t_k$ $(k = 3, \ldots, 11)$. It is concluded that

(i) for $d_l \leq 1.90$: if, for a certain value of d_l, there appears a concave region in the spectral amplitude curve SA_{k-1} at $t = t_{k1}$ then the energy is sufficient to form the k-th soliton;

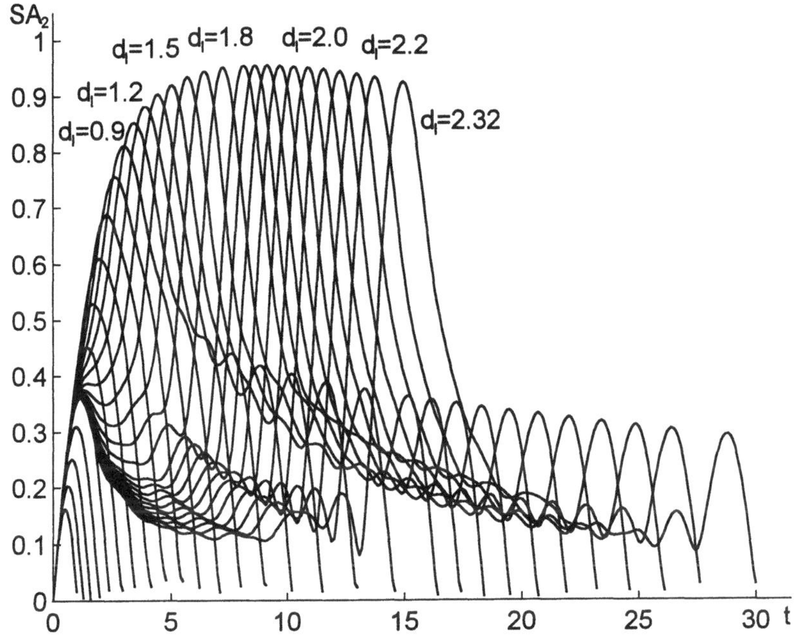

Fig. 5.19. Spectral density S_2 vrs time t for different values of the parameter d_l.

(ii) for $d_l > 1.90$: the appearance of a concave region at $t = t_{k1}$ in the spectral amplitude curve SA_{k-1} means that for this value of d_l a minimum of k solitons must emerge.

In order to understand the changes in spectral densities S_i and the cumulative spectrum SC_i, the numerical data are collected and analysed (Salupere et al., 1994). The sets of $S_i(t_{i1}), SC_i, SC_{i+1}$ dependent on the value of d_l, permit us to conclude that

(i) n solitons form if the quantity $SC_n > 0.2$ (for $n > 3$);

(ii) the higher the number n of emerging solitons, the more energy is embedded in S_i, $i > n$, i.e. for predicting the number of emerging soliton, not only the value of the single spectral density S_i at $t = t_{i1}$ must be considered but also the cumulative spectrum SC_i.

Soliton formation process. The classical papers (Zabusky and Kruskal, 1965; Zabusky, 1981) have explained this physical process in detail. Some comments are still in order based on current results (Salupere et al., 1994).

The lower harmonics fulfil Zabusky's (1969) rule: the recurrence time t_R, which in our notation is t_{10} is n times the time for the $1/n$ harmonic to reach its first maximum. However, for higher harmonics there is a systematic discrepancy leading to smaller value of t_{10} against the direct calculations. For $d_l = 2.3209$ the estimate from t_{121} is 18% lower than from t_2, according to Fig. 1 of Zabusky (1969), the estimate from t_{71} is about 20% lower than from t_2. This situation is obviously caused by the phase shifts of interacting solitons: the larger ones are accelerated by interactions and the recurrence is shifted in time. Abe and Satofuka (1981) have shown how indistinct the recurrence actually is. Our results have shown that the recurrence is better in the spectral sense rather than in wave profiles. The main role in the "partial" recurrence is played by the odd spectral densities $S_3, S_5, \ldots$ (Fig. 5.16). Shifts in these harmonics cause the crucial changes in the wave profile. Notice that our results are based on the pseudospectral method where all spectral characteristics are recalculated at every time-step. If the calculations are based on Fourier components (Abe and Satofuka, 1981), then the recurrence time rule works without any exception, but the recurrence itself is certainly indistinct as already mentioned above.

Closing remarks. The quest for solitons is still a challenge. The deepness of solitonics is easily understood by the flux of new problems. One of new ideas is to deal with smaller number of modes. Osborne (1993) has proposed to use hyperelliptic functions $\mu_j(x,t)$ as nonlinear Fourier series to be superposed linearly. However, while the soliton interactions are nonlinear, there are additional difficulties in computing as well as in interpreting the results. On the other hand, Osborne (1993) indicates also to the existence of the KdV mode amplitudes in the experiments which are constants in space and time contrary to Fourier modes. This seems to be an interesting direction in solitonics.

Our analysis is based on numerical calculations using the pseudospectral method (Salupere et al., 1994; Engelbrecht et al., 1995). The number of emerging solitons is also calculated from the IST but most of the conclusions are based on observations. Summed up, the conclusions are the following:

– solitons are more and more visible after several interactions;

– if n solitons emerge then the soliton formation process is finished at time when the n-th spectral density has reached its first (global) maximum;

– concavity of spectral amplitude SA_{k-1} means that the energy is sufficient to form k solitons; this condition is strictly satisfied for $k \leq 7$ and conditionally for $k > 7$ (in a sense of a minimum number);

– n solitons form if the cumulative spectral measure $SC_n > 0.2$ $(n > 3)$;

– the number of emerging solitons from the amplitude curves determines all possible solitons including virtual solitons.

5.3.4. Other solitonic systems

Besides the KdV equation and its generalizations, other equations also possess solitary-type solutions. Not pretending to give an overwhelming description, the basic equations should still be presented in order to draw some parallels in forthcoming analysis.

The nonlinear Schrödinger (NLS) equation, describing waves in plasma, in superconductors, in optical fibres, etc. (Dodd et al., 1982) is written in the following form

$$iu_\tau + u_{\xi\xi} + \mid u^2 \mid u = 0. \tag{5.144}$$

This equation has the following solution

$$u = A\,\mathrm{sech}[A/\sqrt{2}(\xi - V_s\tau - \xi_0)]\exp[i(1/2V_s\tau - \Omega\tau + \vartheta)]\,, \tag{5.145}$$

$$\Omega = \frac{1}{2}V_s^2 - \frac{1}{2}A^2\,. \tag{5.146}$$

This solution has four independent parameters A, V_s, ξ_0 and ϑ and is called the "envelope soliton". It is easily seen that the envelope has the form of a hyperbolic secant modulating a monochromatic carrier wave. Solution (5.145) obeys Def. 4, but not the condition of Def. 5, therefore we state

D e f i n i t i o n 9: *Velocity of a NLS soliton does not depend upon the amplitude.*

In physical terms, a KdV soliton is called a low-frequency soliton and, correspondingly, a NLS soliton—a high-frequency soliton.

Another important physical situation incorporates wave motion with a special potential. In a unified theory of elementary particles, dislocations in crystals, magnetic flows, etc., the model equation is the Sine-Gordon (SG) equation (see Dodd et al., 1982, and the references therein)

$$u_{xx} - u_{tt} = \sin u\,. \tag{5.147}$$

This is not an evolution equation but through the transformation

$$\xi_1 = \tfrac{1}{2}(x+t)\,, \quad \xi_2 = \tfrac{1}{2}(x-t)\,, \tag{5.148}$$

Eq. (5.147) yields

$$u_{\xi_1\xi_2} = \sin u\,, \tag{5.149}$$

or

$$(u_{\xi_1})_{\xi_2} = \sin u\,. \tag{5.150}$$

Equation (5.147) has the solution (Bullough and Caudrey, 1980)

$$u = 4\arctan\exp\left[\pm\frac{x - x_0 - V_s t}{(1 - V_s^2)^{1/2}}\right]\,, \tag{5.151}$$

which obeys Def. 4 but not Def. 3, i.e. solution (5.151) is not a solitary pulse wave. Therefore we need

D e f i n i t i o n 10: *SG solitons are either kinks or antikinks the velocities of which do not depend upon the amplitude.*

Note that the SG soliton obeys Def. 2. Indeed, if "+" is retained in (5.151) then the kink satisfies $u \to 0$ for $x \to -\infty$ and $u \to 2\pi$ for $x \to +\infty$. Next, if "−" is retained then $u \to 2\pi$ for $x \to -\infty$ and $u \to 0$ for $x \to +\infty$. However, expression (5.151) yields

$$u_x = \pm 2(1 - V_s)^{-1/2}\mathrm{sech}[(x - x \ - V_s t)(1 - V_s^2)^{-1/2}]\,, \tag{5.152}$$

which is similar to a KdV soliton.

For a more complete list of model situations besides KdV, NLS and SG solitons, the reader is referred to other sources (Lamb, 1980, etc.). Here we represent only a "counterexample". The Phi-Four (φ^4) equation known in the quantum field theory (Bullough and Caudrey, 1980) is

$$u_{xx} - u_{tt} = \pm(u - u^3)\,, \tag{5.153}$$

where the usual notation φ for a field variable is changed to u. The solution to Eq. (5.153) is found in the form of a solitary wave

$$u = \pm\text{sech}[1/\sqrt{2}(x - V_s t)(1 - V_s^2)^{-1/2}] \tag{5.154}$$

corresponding to "+" in Eq. (5.153) or in the form of a kink

$$u = \pm\text{th}[1/\sqrt{2}(x - V_s t)(1 - V_s^2)^{-1/2}] \tag{5.155}$$

corresponding to "−" in Eq. (5.153). Neither of the solutions (5.154) and (5.155) obeys Def. 4, i.e. despite describing a "bell-shape" profile (5.154) or a kink (5.155) they cannot interact with each other without loosing their identity.

Consequently, the type of the profile (Def. 3) is not sufficient for deciding whether the wave is a soliton or not.

The examples given above are not exhaustive and describe only the basic cases. More detailed descriptions of solitary waves including breathers, boomerons, fluxons, nanopterons, and other "exotic" structures may be found elsewhere (see, for example Bishop and Schneider, 1978; Bullough and Caudrey, 1980; Boyd, 1990). In terms of equations one should include the Hirota, Boussinesq, Born-Infeld equations and many others (Bhatnagar, 1979). Nevertheless, the consecutive scheme Def. 1 → Def. 2 → Def. 3 → Def. 4 holds, and the physical situation must be fully described in order to establish definitions on the same level as Defs. 5–7. This scheme is based on the properties of the solutions.

Last but not least one must point out an interesting generalization of the KdV equation. The soliton as a solution to the KdV equation is highly localized in space, nevertheless $\text{sech}^2\xi$ has an infinite span in $-\infty < \xi < +\infty$. This has encouraged the quest for a solitary wave which really is compact in space, i.e. has a finite span. Rosenau and Hyman (1993) have proposed a generalization of the standard KdV equation

$$u_t + (u^2)_\xi + u_{\xi\xi\xi} = 0 \tag{5.156}$$

to a family of equations

$$u_t + (u^m)_\xi + (u^n)_{\xi\xi\xi} = 0 \tag{5.157}$$

with $m > 0$, $1 < n \leq 3$. Denoting a member of that family by $K(m, n)$, it is easy to recall the standard KdV equation as $K(2, 1)$.

Rosenau and Hyman (1993) have shown that the $K(2, 2)$ has a solution

$$u_c(\xi, \tau) = \frac{4\lambda}{3}\cos^2[(\xi - \lambda\tau)/4], \quad |\, \xi - \lambda\tau \,| \leq 2\pi \tag{5.158}$$

and $u_c = 0$ otherwise. They call this solution compacton. The speed of a compacton depends upon its height, while its width is independent of its speed.

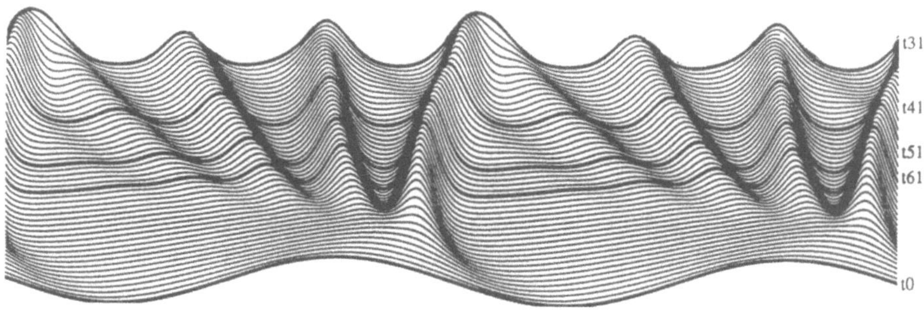

6. Selected case studies

One of the objectives of scientific analysis is to arrange its findings into some patterns.
Encyclopedia Britannica (1963), Science

6.1. DEFORMATION WAVES IN LAYERS WITH MICROSTRUCTURE

6.1.1. Statement of the problem

Wave propagation in layered elastic media has been intensively studied for a long time (Ewing et al., 1957) and the keyword has been dispersion. Contemporary solid mechanics as described above needs another essential keyword to be added—that is nonlinearity. In addition there may be cases when the layers possess microstructure (see Section 5.1). There are several attempts to describe waves in such layers (Engelbrecht and Khamidullin, 1988; Engelbrecht, 1991; Engelbrecht and Peipman, 1992). Here we present their ideas in order to demonstrate a phenomenological approach for modelling complicated processes based on analogies and asymptotics for deriving evolution equations.

The need for such a complicated mathematical problem has arisen in seismology (Nikolaev, 1989). The crustal zone as a geophysical medium may, beside its other properties, be characterized by locked-in internal energy. For example, microseismic oscillations (microseisms) are constantly monitored and usually explained in terms of seismic emission. The detailed analysis of the spectra of microseisms shows that they cannot be generated solely by external sources, and the role of internal sources must also be crucial. Also it is assumed that the external sources may act as a trigger mechanism, resulting in the release of the locked-in internal energy. If a propagating seismic wave is considered to be an external source, then this main wave itself may be amplified due to additional energy influx. Recent experimental data show explicitly the role of "activity" in a geophysical medium.

Within the standard theory of elasticity the influence of such an "activity" can not be described. Here we assume that it is caused by the microstructure and try to derive an evolution equation governing waves in a horizontal layer. Contrary to the classical theory where layers are considered to be homogeneous, it is assumed that the Earth's crust is made up by a certain hierarchy of blocks which are connected by thin interface layers (Sadovski, 1985). These layers are interpreted as inhomogeneities which are the most probable regions of fracture due to shear or shear-tensile stresses (Kozak and Šilený, 1985). Let us consider wave motion in the layer of such a block structure. For oceanic paths, for example, the thickness h of the Earth crust is about 5 km and the usual wave length is approximately $10^2 \dots 10^3\ h$, i.e. two or three orders higher than the thickness of the crust. The assumption about long waves means that despite of complicated (and random) structure of the layer, it can be modelled as a homogeneous object with a certain microstructure. The main hypothesis in the modelling is the following: the behaviour of a wave depends upon its intensity and upon the properties of the microstructure. A low-intensity wave creates additional stress in the interface layers and so a residual stress field is created serving as a prestress field for next waves. In other words, energy from the wave is pumped into the structure and the wave itself is attenuated. The stresses grow with increasing intensity of waves until they reach critical value. Any high-intensity wave may release this energy embedded into the structure. Physically it means that either the fracture of the interface layers occurs or the blocks simply slide with respect to each other. Energy released in such a manner may amplify the main wave which has triggered the energy release. This hypothesis will later be explained in more details.

6.1.2. Physical description

In order to understand the mechanism of possible accumulation and release of energy, the attention will first be focused on fracture mechanics. According to the theory of microcrack formation related to the deformation energy, an interesting concept is proposed by Zhurkov (1983) and Petrov (1983). This concept starts from the assumption that the negative density fluctuations of a medium with loosened bonds between the atoms cause fluctuations of the internal energy. Such fluctuations are called dilatons, i.e. dilaton is a short-lived microdynamical density fluctuation which is able to absorb energy from the surrounding medium because the bonds between atoms in a dilaton may be loosened. Since thermodynamical considerations do not allow an infinite increase of the internal energy, a certain threshold must exist. When the accumulated energy in the dilaton has reached the critical value, the dilaton is no longer able to exist and it breaks up releasing the stored energy. The breaking-up means that a certain microcrack has been formed. The cumulative process of the formation of microcracks leads finally to fracture (Zhurkov, 1983).

The quantitative estimations about the dilaton mechanism include the limit value of stress σ_d and deformation ε_d (Zhurkov, 1983) as well as the activation energy (Petrov, 1983). For example, the limit deformation is given by the formula

$$\varepsilon_d = \frac{\alpha T}{3} ln \frac{\tau}{\tau_0} \tag{6.1}$$

and the stress, respectively, by the formula

$$\sigma_d = E\varepsilon_* \left(1 - \frac{\varepsilon_d}{\varepsilon_*}\right) . \tag{6.2}$$

Here the following notations have been used:

E is the modulus of elasticity, T is the absolute temperature, $\varepsilon_*(\approx 0.2)$ is the limit deformation for the bonds between the atoms, τ is fracture time, τ_0 is the time constant at the order of atom oscillations ($\sim 10^{-13}s$), and α is the coefficient of thermal expansion. A dilaton breaks up fast ($\tau_d \sim 10^{-9}s$) and the characteristic scale for a microcrack (a crack nucleus) formed as a result of this process is about 10^{-5} cm. The limit values of σ_d and ε_d for several metals and haloid crystals are given by Zhurkov (1983) at room temperature and at $\tau = 1s$ (i.e. $ln(\tau/\tau_0) \sim 30$).

It is clear that this concept can not be used directly in the theory of seismic waves because the structural elements (and the wave lengths) in fracture mechanics and in seismology differ by many orders of magnitude. Nevertheless, the basic idea of a dilaton can also be generalised for larger structural elements. In order to make clear difference between the dilatons of Zhurkov (1983) and Petrov (1983) and dilatons in seismology, we call them micro- and macrodilatons, respectively.

Before the concept of macrodilatons of a block structure is formulated, several comments are needed. The idea of seismic energy release is, for example, presented by Kozak and Šilený (1985) and Gusev (1988). We follow here Kozak and Šilený (1985) who considered a two-dimensional homogeneous space containing a linear inhomogeneity. The space is subjected to unaxial tectonic stress. The process is explained in the following way: "The relations among the treated inhomogeneity, the tectonic stresses loading it and a possible resulting displacement resemble a family story: the "maternal" medium comprising the inhomogeneity—the stress concentrator—is subjected to the acting of the "paternal" tectonic stress... With increasing load, the size of the zone affected by the tectonic stress concentrations grows until the stress field components reach their critical values at the points of their maximum concentrations. A seismically active (seismoactive, or seismic-energy-releasing) event of the shear or combination shear-tensile type can be born..." (Kozak and Šilený, 1985). It is easily concluded that such a situation can also be classified as a dilaton (macrodilaton) mechanism. In the experiments by Kozak and Šilený (1985), the stress concentrator in a square model was simulated by an open linear fault inclined with respect to the direction of unaxial compression.

The new-born tensile cracks had lengths of the order of tens of mm and the parameters of the seismic energy release (amplitudes and frequencies) did not depend on the inclination of the fault. Moreover, the energy release was found to be associated only with the initial stage of tensile cracks. A more detailed analysis of energy outflow from a growing crack is given by Slepyan (1990) with an important conclusion—the structural properties of a medium should be accounted for in the theory in order to solve crack formation problems properly.

Thus, dealing with waves in the Earth's crust, the following simplified physical model is proposed. The waves propagate in a layer resting on a halfspace. The layer is assumed to have a block structure (see Fig. 6.1). Two effects must be accounted for: (i) the influence of this block structure and (ii) dispersion in the layer. The dispersion relation, for example, at SH waves in an elastic layer is known (Ewing et al., 1957). The influence of the block structure is taken into account using the concept of macrodilatons based on seismic description. Schematically, seismic macrodilatons are shown in Fig. 6.2. Based on this scheme and on the general concept of dilatons, the following working hypothesis is proposed (Engelbrecht and Khamidullin, 1988): low-intensity seismic waves give a part of their energy to dilatons (Fig. 6.2b) while the high-intensity (above a certain threshold) seismic waves cause the dilatons to break up (Fig. 6.2c). This means that in the first case the dilaton mechanism causes additional attenuation of the propagating wave, and in the second case the released internal energy (caused by the developing of a system of microcracks) forms an energy influx with respect to the propagating wave resulting in amplification.

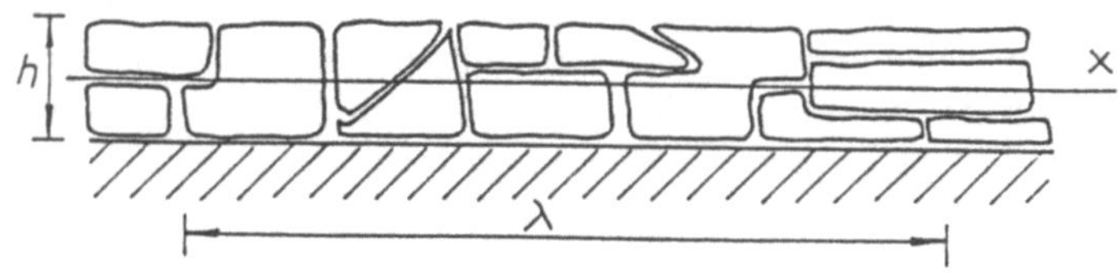

Fig. 6.1. Sketch of a layer with a block structure on a halfspace (hatched region); h – thickness, λ – characteristic wavelength.

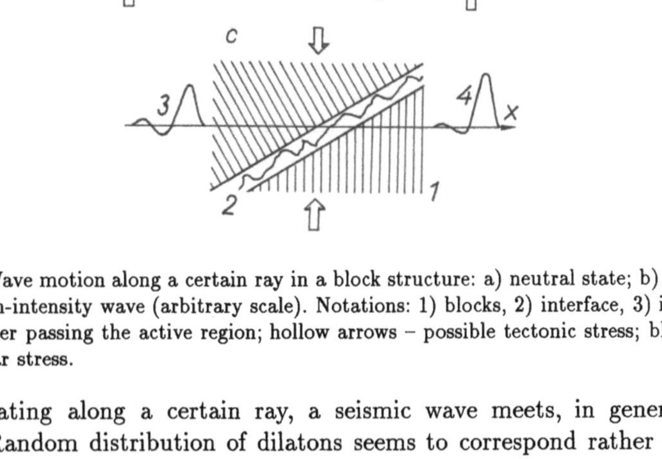

Fig. 6.2. Wave motion along a certain ray in a block structure: a) neutral state; b) low-intensity wave; c) high-intensity wave (arbitrary scale). Notations: 1) blocks, 2) interface, 3) initial profile, 4) profile after passing the active region; hollow arrows – possible tectonic stress; black arrows – possible shear stress.

Propagating along a certain ray, a seismic wave meets, in general, various dilatons. Random distribution of dilatons seems to correspond rather well to the real situation, but a model problem of homogeneous dilatons can also be of interest in order to understand the main character and parameters of the process.

6.1.3. *Phenomenological description*

After the working hypothesis is fixed (see Section 6.1.2), a mathematical model governing wave motion in a layer should be constructed. Here a simplified model accounting for the homogeneous (regular) distributions of dilatons will be presented.

The governing equations, as usual, are described by conservation laws and constitutive equations. We use the balance of momentum

$$(T^{KL}x^k_{,L})_{,K} + \rho_0(f^k - A^k) = 0\,, \qquad (6.3)$$

where the notations of Chapter 2 are used. To get a closed system, the constitutive equations must be postulated for both T^{KL} and f^k, i.e. we need additional postulates. Based on thermodynamic consideration, we might also use the energy balance, involving heat supply and, consequently, temperature T (see Chapters 2 and 4). First, the constitutive law

$$T^{KL} = T^{KL}(E_{KL}, \dot{E}_{KL}, T, \ldots) \qquad (6.4)$$

must be stated. If $f_k = 0$ then the system (1)–(4) is closed and the variables U_K, T are determined under given initial and boundary conditions. According to our working hypothesis, the dilaton mechanism leads to the nonmonotonic character of the process that must be taken into account when formulating a constitutive law. Obviously, when considering high- and low-intensity waves, our model must be nonlinear in order to describe properly such a motion.

The qualitative analysis of (1)–(4) may be summarized briefly:

(1) let us assume that physical (nonlinear function (4)) and geometrical (finite deformation) nonlinearities are taken into account. Such a model describes the interaction of waves as well as the generation of higher harmonics (Bland, 1969; Engelbrecht, 1983). The energy function is usually monotonic.

(2) Let us assume that dissipation is taken into account. Linear dissipation (Ricker, 1977) leads to attenuation which is proportional to either the square or the square root of frequency, for both low and high frequencies, respectively. Nonlinear dissipation (Eringen, 1962) may essentially change the energy absorption but under the rules of material frame indifference and the principle of equipresence, many terms of the same order need to be taken into account. As a result, the phase velocity depends on the deformation rate, and a possible explanation of the dilaton mechanism is not explicit, at least for a model problem. A similar situation arises when the attenuation coefficients depend on the deformation.

(3) Let us assume that several thermodynamic parameters exist, being measures of other physical fields beside the mechanical deformation. These may be, for example, an electromagnetic or temperature field, or the thermodynamic parameters (inter-nal variables) that may phenomenologically describe the effect of chemical reactions. Since these parameters come into account through energetic relations, then temperature effects as well as the influence of coupling are important. Such a situation arises in the near vicinity of the epicentre of an earthquake, and therefore may be more important for large-scale fluctuations.

Motivated by this brief analysis we find a very simple mathematical model: excluding other possible effects, there remains only one physically admissible and mathematically simple possibility—the body forces f_k which we shall assume to be related to the dilaton mechanism. In other words, another constitutive law

$$f_k = f_k(E_{KL}) \tag{6.5}$$

must be postulated to close the system (6.3), (6.4). Physically this assumption is transparent—the body forces result from long-range effects (Eringen, 1962) and many examples are known with body forces taken into account. For example, the motion of a thin liquid layer on a sloping surface requires that besides the capillary forces also the weight of the moving liquid is taken into account. Many examples exist in electronics where driving forces cause qualitative changes in output parameters (Scott, 1970).

6.1.4. Mathematical model

Let us consider the temperature changes in propagating seismic waves to be so small that they may be neglected. Then the waves are governed by the balance of momentum (6.3) together with constitutive laws (6.4) and (6.5).

At this stage, for the example being explained, we specify the type of waves being horizontally polarized shear (SH) waves. For the one-dimensional case the equation of motion (6.3) yields

$$\frac{\partial^2 U_2}{\partial t^2} = c_s^2 \frac{\partial^2 U_2}{\partial X_1^2} + c_s^2 m_1 \left(\frac{\partial U_2}{\partial X_1}\right)^2 \frac{\partial^2 U_2}{\partial X_1^2} + f_2 \,, \tag{6.6}$$

where c_s is the shear wave velocity and m_1 is the nonlinear parameter involving the moduli of the second, third and fourth orders (Bataille and Lund, 1982). The corresponding evolution equation describing the wave motion along the positive direction of X_1 is derived by standard methods (see Chapter 4)

$$\frac{\partial u}{\partial \tau} - \frac{m_1}{2\varepsilon^2 c_s^2} u^2 \frac{\partial u}{\partial \xi} + \frac{1}{2\varepsilon c_2} f_2 = 0 \,, \tag{6.7}$$

where $\xi = c_2 t - X_1$, $\tau = \varepsilon^2 X_1$, $u = \partial U_2/\partial t$ and ε is a small parameter related to the maximal deformation gradient. We shall assume that f_2 may be expressed in terms of the gradient $U_{2,1}$

$$f_2 = -(b_1 U_{2,1} + b_2 U_{2,1}^2 + b_3 U_{2,1}^3) \,. \tag{6.8}$$

Here b_1, b_2, b_3 are positive constants. As $\partial U_2/\partial t = -c_2 U_{2,1}$ in the first approximation this can be rewritten as

$$f_2 = B_1 u - B_2 u^2 + B_3 u^3 \,, \tag{6.9}$$

where $B_1 = b_1/c_s$, $B_2 = b_2/c_s^2$, $B_3 = b_3/c_s^3$ are also positive.

The dispersion relation for horizontally polarized shear waves is given by Ewing et al. (1957)

$$\tan K_1 h = (\mu_2 K_2)/(\mu_1 K_1) \,, \tag{6.10}$$

$$K_1 = (\kappa^2 - \omega^2/c_{s1}^2)^{1/2} \,, \tag{6.11}$$

$$K_2 = (\omega^2/c_{s2}^2 - \kappa^2)^{1/2} \,. \tag{6.12}$$

Here subscripts 1 and 2 refer to the layer and halfspace, respectively; μ_i are the Poisson ratios and ω, κ are the frequency and wave number, respectively.

It has been shown that for long waves ($\kappa h \ll 1$) the dispersion relation (6.10) may be simplified to (Bataille and Lund, 1982)

$$\omega^2 = c_{s2}^2(\kappa^2 - l_0^2\kappa^4)\,, \tag{6.13}$$

$$l_0^2 = h^2(c_{s2}^2/c_{s1}^2 - 1)^2\mu_1^2/\mu_2^2 > 0\,. \tag{6.14}$$

In this case the governing equation of motion can be re-established in the form (Bataille and Lund, 1982)

$$\frac{\partial^2 U_2}{\partial t^2} = c_{s2}^2\frac{\partial^2 U_2}{\partial X_1^2} + c_{s2}^2 l_0^2\frac{\partial^4 U_2}{\partial X_1^4}\,, \tag{6.15}$$

which is correct for long waves only.

Following Whitham (1974) we combine (6.6) and (6.15) arriving at

$$\frac{\partial^2 U}{\partial t^2} = c_s^2\frac{\partial^2 U_2}{\partial X_1} + c_s^2 m_1\left(\frac{\partial U_2}{\partial X_1^2}\right)^2\frac{\partial^2 U}{\partial X_1^2} + c_s^2 l_0^2\frac{\partial^4 U_2}{\partial X_1^4} + f_2\,, \tag{6.16}$$

where c_s is meant as c_{s2}. The corresponding evolution equation is now

$$\frac{\partial u}{\partial \tau} - \frac{m_1}{2\varepsilon^2 c_s^2}u^2\frac{\partial u}{\partial \xi} + \frac{l_0^2}{2\varepsilon}\frac{\partial^3 u}{\partial \xi^3} + \frac{1}{2\varepsilon c_s}f_2 = 0\,, \tag{6.17}$$

which describes the propagation of long nonlinear SH waves in a layer with a driving force, i.e. with possible energy influx.

Introducing dimensionless variables by the expressions

$$\bar{u} = u/u_0\,, \quad \bar{\tau} = k_1 u_0^2\tau\tau_c^{-1}\,, \quad \bar{\xi} = \xi\tau_c^{-1} \tag{6.18}$$

where

$$k = \frac{1}{2}\mid m_1 \mid \varepsilon^{-2}c_s^2 \tag{6.19}$$

we get (bars over dimensionless variables are omitted)

$$\frac{\partial u}{\partial \tau} - \operatorname{sgn}\mid m_1 \mid u^2\frac{\partial u}{\partial \xi} + \nu\frac{\partial^3 u}{\partial \xi^3} = -(\beta_1 u - \beta_2 u^2 + \beta_3 u^3)\,. \tag{6.20}$$

Here

$$\nu = \varepsilon l_0^2 c_s^2(\tau_c^2 u_0^2 \mid m_1 \mid)^{-1}\,, \quad \beta_1 = QB_1/u_0^2\,, \quad \beta_2 = QB_2/u_0\,,$$
$$\beta_3 = QB_3\,, \quad Q = \tau_c\varepsilon c_s \mid m_1 \mid^{-1}\,, \tag{6.21}$$

while u_0 is the initial amplitude of the particle velocity and τ_c is the effective wavelength.

The outcome for SH waves is the modified KdV equation with a r.h.s. For longitudinal (P) waves (Naugolnykh and Ostrovsky, 1990), the nonlinearity has a quadratic character yielding in a term $u\partial u/\partial\xi$ (in suitable chosen variables). The main problem for both longitudinal and transverse waves is the integration of a KdV-type equation with a r.h.s. which here is approximated by a cubic polynomial.

6.1.5. Numerical simulation

We use the PsM (see Section 5.3) for integrating a KdV-type equation with a r.h.s. In terms of Section 5.3 we have

$$u_t + uu_x + du_{xxx} = \varepsilon f(u) \tag{6.22}$$

$$f(u) = -(a_1 u - a_2 u^2 + a_3 u^3) \tag{6.23}$$

where a_1, a_2, a_3 are positive constants and ε is a small parameter (different from the small parameter used earlier). Again, as in Section 5.3, we have used t, x for τ, ξ.

With $\varepsilon = 0$, Eq. (6.22) has a soliton-type solution. When the initial excitation is in the form of a single soliton, then this should propagate without any changes in its shape. Indeed, in Fig. 6.3 the evolution of a single soliton is shown and demonstrates this property. The soliton is shifted with respect to $\xi = 0$ (Fig. 6.3a) and the spectral amplitudes SA_k, as supposed, remain constant (Fig. 6.3b). If $\varepsilon \neq 0$, then essential changes occur in the shape of the soliton. We take here $a_1 = 1, a_2 = 0.5, a_3 = 0.0556$ that are chosen for illustrative purpose only. Figure 6.4 shows the evolution in this case. The slight amplification occurs but the soliton gets a new stable shape. This is easily seen from the behaviour of spectral amplitudes (Fig. 6.4b) that after a certain transition period reach new constant values. A typical feature of this stable shape is asymmetric with respect to the central axis (i.e. with respect to the coordinate of the maximum amplitude). For smaller amplitudes of the initial soliton, the attenuation is expected, as is also shown in Fig. 6.5.

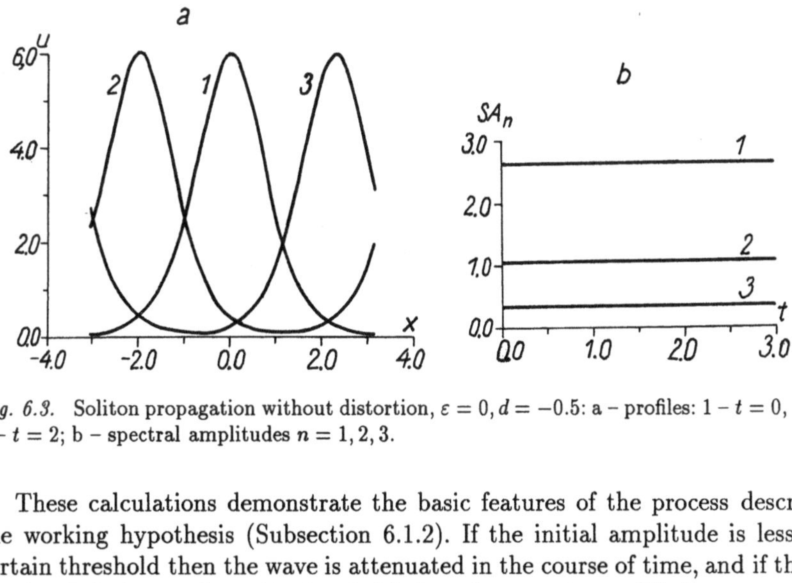

Fig. 6.3. Soliton propagation without distortion, $\varepsilon = 0, d = -0.5$: a – profiles: $1 - t = 0$, $2 - t = 1$, $3 - t = 2$; b – spectral amplitudes $n = 1, 2, 3$.

These calculations demonstrate the basic features of the process described by the working hypothesis (Subsection 6.1.2). If the initial amplitude is less than a certain threshold then the wave is attenuated in the course of time, and if the initial amplitude is larger than this threshold then the wave is amplified.

The soliton formation from periodic initial conditions under the influence of a r.h.s. may be closer to the conditions of real world than the pure conservative case. The latter was throughly examined in Section 5.3. Characteristic to this process is the formation of soliton trains that are in constant interaction process.

Suppose now that a r.h.s. exists, i.e. the solitons are formed in a medium with microstructures. Then the threshold dividing the attenuation/amplification regions can be such that the larger solitons undergo amplification and the smaller ones—attenuation. The full picture should then be completely different from what was described in Section 5.3. Indeed, Fig. 6.6 demonstrates such a situation. For r.h.s. equal to zero, i.e. for a conservative case, three solitons are formed (Fig. 6.6a). The

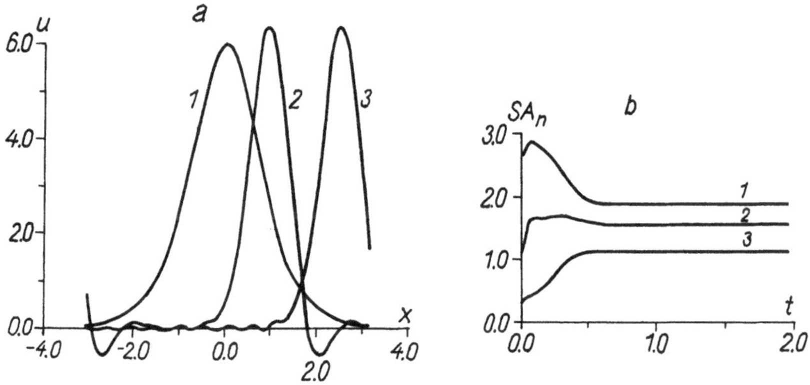

Fig. 6.4. Soliton propagation with amplification, $\varepsilon \neq 0, d = -0.5$: a – profiles: $1 - t = 0$, $2 - t = 1$, $3 - t = 2$; b – spectral amplitudes $n = 1, 2, 3$.

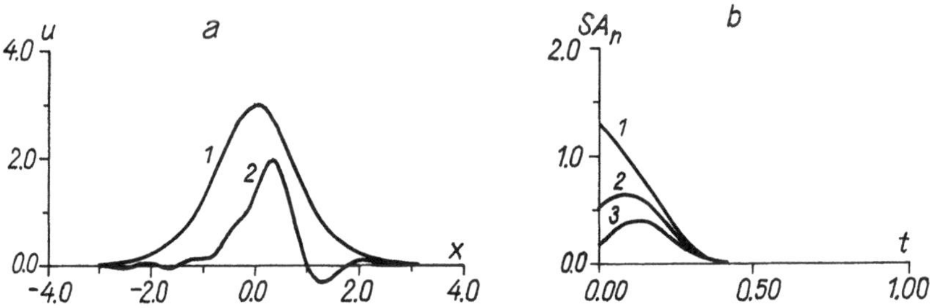

Fig. 6.5. Soliton propagation with attenuation, $\varepsilon \neq 0, d = -0.5$: a – profiles: $1 - t = 0$, $2 - t = 0.2$; b – spectral amplitudes $n = 1, 2, 3$.

case of $\varepsilon f(u) \neq 0$ leads to the survival of only the largest of them (Fig. 6.6b). This soliton is stable as seen from the spectral amplitudes.

Returning now to the problem of seismic waves stated above the proposed phenomenological model indeed describes the possible amplification or attenuation of waves. Moreover, it includes a certain selective transition period during which only high-intensity waves survive while the smaller waves will be attenuated. This can be the case in formation of precursors of long high-intensity seismic waves. The physical properties of a precursor as a stable solitary wave are analysed by Lund (1983) for the Great Chilean Earthquake in 1960.

6.2. DEFORMATION WAVES WITH THERMAL EFFECTS

6.2.1. Statement of the problem

Every disturbance, i.e. wave in the physical world propagates with a finite velocity. In mathematical terms it means the hyperbolicity of the governing equations. As

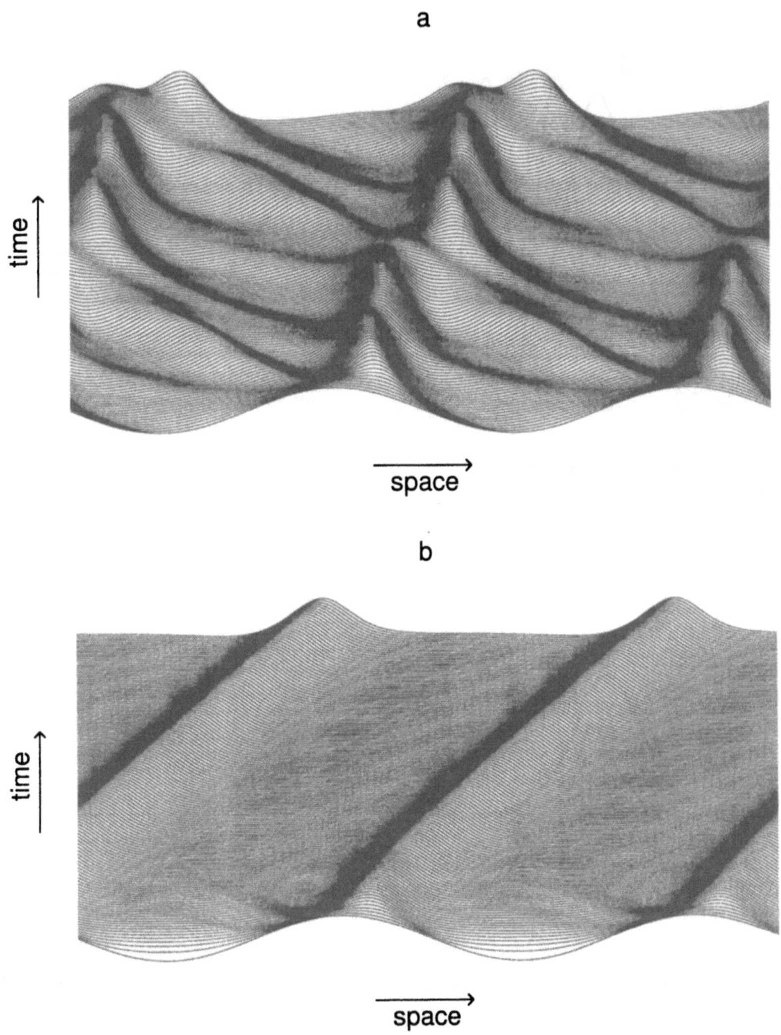

Fig. 6.6. Time slice plots for unperturbed (a) and perturbed (b) KdV equation for $d_l = 1.1$; $P(u) = \varepsilon f(u) = -(1.125u - 2.25u^2 + u^3)$.

already discussed in previous chapters, the reality leads to such complications in mathematical models that various asymptotic methods have been used or simplifying assumptions have been introduced for deriving the governing equations so that the outcome is not necessarily hyperbolic. A very general case is demonstrated in Chapter 4 when explaining the idea of evolution equations. These are not always hyperbolic but there is a characteristic velocity, i.e. a moving frame involved, and the initial mathematical model and the evolution equations are related through an asymptotic scheme. As shown in Section 4.2, in this scheme one must distinguish an associated system which is always hyperbolic.

The problem is much more complicated when we have to deal with a mixed system or there are difficulties in extracting an associated system. The example, described below, deals with the case of thermoelasticity in the 1D approach. This is a typical hyperbolic-parabolic problem.

It is well-known that the Fourier law states

$$Q_1 = kT_{,1} \tag{6.24}$$

that leads to the parabolic heat conduction equation with the infinite velocity of propagation of heat. There has been a long discussion about the proper form of governing equations in modified theories in order to remove this paradox (see, for example, the review by Chandrasekharaiah, 1986). One of the possibilities to overcome that is to introduce the relaxation time τ_0 of the heat flux and to use instead of (6.24) the modified law

$$\tau_0 \dot{Q}_1 + Q_1 = kT_{,1}\,. \tag{6.25}$$

This idea actually dates back to Maxwell but is more known as the Vernotte-Cattaneo model. The modified rate-type law of heat conduction (6.25) leads to the hyperbolic heat conduction equation and then seemingly the mathematical models are paradox-free. However it is known (Engelbrecht, 1983) that the influence of the relaxation time τ_0 in thermoelasticity is essential for high-frequency processes mostly (shocks or oscillations in the order of the characteristic velocity). On the other hand, the relaxation mechanism is well understood on the basis of phonon-phonon interaction for dielectrics (Landau, 1941). In metals, where the electron mechanism is more essential, the results obtained by various authors differ greatly (Francis, 1972). It seems that the theories based on more sophisticated arguments than the improved heat conduction law (6.25) may better remove the paradox of infinite speed (Chandrasekharaiah, 1990; Lebon et al., 1992).

In thermoelasticity, the equation of motion of deformation waves is coupled with the heat conduction equation. If Eq. (6.25) is used then we have to deal with a system from which a hyperbolic associated system is easily deduced. If Eq. (6.24) is used then the outcome is a mixed hyperbolic-parabolic system. Here we analyse both cases in order to demonstrate various approaches described above for the analysis of complicated wave processes. The first is based on using the modified heat conduction law (6.25) and the straight-forward derivation of evolution equations (Section 4.2). The second is based on using the classical heat conduction law (6.24) and the concept of internal variables (Section 4.3). In both cases we derive the evolution equations for deformation waves but they are different because of the different assumptions introduced at various stages of analysis. As a result, the thermal attenuation is accounted for by different terms and the outcomes could be characterized as "equilibrium" and "instantaneous" cases.

6.2.2. Analysis with finite velocities

Based on models of Section 4.1, we start with

$$(\lambda + 2\mu)[1 + 3(1 + m_0)U_{1,1} - \kappa(\lambda + 2\mu)^{-1}(T - T_0)]U_{1,11} - $$
$$- \kappa(1 + U_{1,1})T_{,1} - \rho_0 \ddot{U}_1 = 0\,, \tag{6.26}$$

$$T\kappa \dot{U}_{1,1} + TT_0^{-1}\rho_0 c_E \dot{T} = Q_{1,1}\,, \tag{6.27}$$

$$\tau_0 \dot{Q}_1 + Q_1 = kT_{,1}\,, \tag{6.28}$$

where the notations are given above (see also Engelbrecht, 1983). We introduce the parameters

$$e = \kappa^2 T_0[(\lambda + 2\mu)\rho_0 c_E]^{-1}\,, \tag{6.29}$$

$$\beta_r = c_0 \tau_0 L^{-1}\,, \tag{6.30}$$

$$w_d = L\rho_0 c_E c_0 k^{-1}\,, \tag{6.31}$$

where L is a specific space measurement, for example wavelength τ_c.

These dimensionless parameters govern the thermoelastic wave process: the parameter e is the standard coupling parameter, the parameter β_r governs relaxation (Lord and Shulman, 1967) and the parameter w_d – diffusion (Johnson, 1975). The dependent and independent variables can also be presented in a dimensionless form:

$$\tilde{X} = X_1\tau_c^{-1}\,, \quad \tilde{t} = c_0 t \tau_c^{-1}\,, \quad \tilde{U}_1 = (\lambda + 2\mu)(\kappa T_0 \tau_c)^{-1} U_1\,,$$
$$P = Q\tau_c(kT_0)^{-1}\,, \quad \theta = (T - T_0)T_0^{-1}\,. \tag{6.32}$$

The tilde over the dimensionless variables is now dropped and the basic system (6.26)–(6.28) is written in the form

$$I\frac{\partial \mathbf{U}}{\partial t} + A\frac{\partial \mathbf{U}}{\partial X} + \mathbf{H} = 0\,, \tag{6.33}$$

$$\mathbf{U} = \begin{vmatrix} \dot{U}_1 \\ U_{1,1} \\ \theta \\ P \end{vmatrix}\,, \qquad \mathbf{H} = \begin{vmatrix} 0 \\ 0 \\ 0 \\ \beta_r^{-1}P \end{vmatrix}\,, \tag{6.34}$$

$$A = \begin{vmatrix} 0 & -1 - 3(1+m_0)\varepsilon U_{1,1} + \varepsilon\theta & 1 + \varepsilon U_{1,1} & 0 \\ -1 & 0 & 0 & 0 \\ e(1 + \varepsilon U_{1,1}) & 0 & 0 & -\omega_d^{-1} \\ 0 & 0 & -\beta_r^{-1} & 0 \end{vmatrix}\,. \tag{6.35}$$

Here $\varepsilon = \kappa T_0(\lambda + 2\mu)^{-1}$. The moving frame is assumed to be

$$\xi = (1+e)^{1/2}t - X\,, \quad \tau = \varepsilon X\,. \tag{6.36}$$

In the first approximation (see Section 4.2)

$$\mathbf{U}_0 = v\begin{vmatrix} -(1+e)^{1/2} \\ 1 \\ e \\ -e\partial/\partial\xi \end{vmatrix}\,, \tag{6.37}$$

where v is governed by the evolution equation (in terms of $u = vv_0^{-1}$)

$$\frac{\partial u}{\partial \tau} + \mathrm{sign}(1 + m_0 + e)u\frac{\partial u}{\partial \xi} = \Gamma^{-1}\frac{\partial^2 u}{\partial \xi^2}\,, \tag{6.38}$$

$$\Gamma = 3(1+e)^{1/2} \mid 1 + m_0 + e \mid v_0 \tau_c \rho_0^2 c_E^2 c_0 (\kappa k)^{-1}\,. \tag{6.39}$$

This is the famous Burgers equation governing the possible shock profiles. What is important, is the fact that the relaxation parameter β_r does not influence the evolution equation. More details are given by Engelbrecht (1983).

6.2.3. Formalism of internal variables

Here we deal with the standard Fourier law (6.24). Instead of the system (6.26)–(6.28) we start with

$$(\lambda + 2\mu)[1 + 3(1 + m_0)U_{1,1}]U_{1,11} - \kappa T_{,1} - \rho_0 \ddot{U}_1 = 0\,, \tag{6.40}$$

$$T_0 \kappa \dot{U}_{1,1} + \rho_0 c_E \dot{T} = Q_{1,1}\,, \tag{6.41}$$

$$Q_1 = kT_{,1}\,. \tag{6.42}$$

Note that some nonlinearities are neglected, too, while they did not affect the results in Section 6.2.2.

Now we follow the formalism of internal variables presented in Section 4.3 (for more details see Engelbrecht and Maugin, 1996) and treat the system (6.40)–(6.42) as an inertial-dissipative system. This means that the displacement U_1 is treated as an observable variable and the temperature T as an internal variable.

The observable variable is governed by the system

$$I\frac{\partial \mathbf{U}}{\partial t} + A\frac{\partial \mathbf{U}}{\partial X_1} + \mathbf{H} = 0\,, \tag{6.43}$$

$$\mathbf{U} = \begin{vmatrix} \dot{U}_1 \\ U_{1,1} \end{vmatrix}, \quad \mathbf{H} = \begin{vmatrix} \kappa \rho_0^{-1} T_{,1} \\ 0 \end{vmatrix}, \tag{6.44}$$

$$A = \begin{vmatrix} 0 & -c_0^2[1 + 3(1 + m_0)U_{1,1}] \\ -1 & 0 \end{vmatrix} \tag{6.45}$$

and the internal variable—by the equation

$$T_0 \kappa \dot{U}_{1,1} + \rho_0 c_E \dot{T} = kT_{,11}\,. \tag{6.46}$$

Applying now the scheme of Section 4.2, and assuming $H \sim O(\varepsilon)$, i.e. small, we get for a wave propagating to the right the evolution equation

$$\frac{\partial u}{\partial \tau} + \frac{3(1 + m_0)}{2\varepsilon c_0} u \frac{\partial u}{\partial \xi} + \frac{\kappa}{2\varepsilon c_0 \rho_0} T_{,1} = 0\,, \tag{6.47}$$

where the moving frame is $\xi = c_0 t - X_1$, $\tau = \varepsilon X_1$, and $u = \dot{U}_{1(0)}$. The small parameter ε, as usual, is related to the initial excitation and in terms of the particle velocity u it means that its maximum value u_0 is much less than the sound velocity c_0, i.e.

$$\varepsilon = \frac{3}{2}(1 + m_0)u_0 c_0^{-1}\,. \tag{6.48}$$

It is easily seen that in (6.47) temperature is a parameter treated as an internal variable. In terms of the moving frame ξ, τ, Eq. (6.46) reads

$$-T_0 \kappa \frac{\partial u}{\partial \xi} + \rho_0 c_E c_0 \frac{\partial T}{\partial \xi} = k \frac{\partial^2 T}{\partial \xi^2}\,, \tag{6.49}$$

where the accuracy is of $O(1)$, i.e. terms of order $O(\varepsilon)$ are neglected. As it is easily seen, the order of Eq. (6.49) can be reduced. Integrating it once, we obtain

$$-T_0 \kappa \alpha + \rho_0 c_E c_0 T = k \frac{\partial T}{\partial \xi} + C\,, \tag{6.50}$$

where C is the constant. Assuming small changes of temperature, we may take $T \sim T_0$. Selecting then

$$C = \rho_0 c_E c_0 T_0\,, \tag{6.51}$$

Eqs. (6.47) and (6.50) yield (after transforming $T_{,1}$ to the moving frame)

$$\frac{\partial u}{\partial \tau} + a_1 u \frac{\partial u}{\partial \xi} + a_2 u = 0 \tag{6.52}$$

with

$$a_1 = \frac{3(1+m_0)}{2\varepsilon c_0}, \quad a_2 = \frac{\kappa^2 T_0}{2\varepsilon k c_0 \rho_0}. \tag{6.53}$$

Scaling dependent and independent variables with τ_c and u_0, respectively, the dimensionless equation is easily obtained (retaining the same notations)

$$\frac{\partial u}{\partial \tau} + \text{sign} \mid 1 + m_0 \mid u \frac{\partial u}{\partial \xi} + \Lambda u = 0\,, \tag{6.54}$$

$$\Lambda = \frac{\kappa^2 T_0 \tau_c}{3 \mid 1 + m_0 \mid k \rho_0 u_0}. \tag{6.55}$$

Equation (6.54) is the sought evolution equation for a deformation wave with thermal attenuation provided the temperature is treated as an "internal" variable. The solution of (6.54) may also include discontinuities when the gradient of the initial excitation is large enough (Whitham, 1974).

The difference between the results of Section 6.2.2 (the Burgers equation) and Section 6.2.3 (the simple wave equation) is easily explained. When the basic model was of the hyperbolic type, the changes in temperature T were estimated as $O(\varepsilon)$. When the basic model was of hyperbolic-parabolic type, the temperature rate $T_{,1}$ was of the $O(\varepsilon)$ order. Consequently, Eq. (6.54) describes the low-frequency losses while Eq. (6.37) corresponds to the high-frequency losses. In this sense, the formalism of the internal variables helped to remove the discrepancy of the traditional approach which permits only the smoothening of an arbitrary excitation. In other words, Eq. (6.54) corresponds to a so-called instantaneous wave for which one focuses on what occurs in the immediate vicinity of the wave front as the latter travels down the thermoelastic material. The viewpoint of Eq. (6.38) corresponds to that of the delayed wave where one is especially interested in a far-field solution at distances of the order ε^{-1} from the source.

6.3. ULTRASONIC WAVES IN TISSUES

6.3.1. The complexity of ultrasonic wave motion

Ultrasound in the kilohertz and megahertz frequency range is widely used in medicine. It is being utilized in many ways (Crum et al., 1986): (i) as a diagnostic instrument, (ii) as a therapeutic device, and (iii) as a surgical tool including "ultrasonic scalpels" and kidney stone disrupters. Recent strides in technological development (particularly in semiconductor circuitry) have revolutionized the capability and sophistication of the instrumentation available for above-mentioned purposes. On the another hand, theoretical research in ultrasonic diagnostics and surgery is often limited by the relatively simple models of propagation invoked in what is largely an empirical science. Careful discussion of these problems reveals the need for more sophisticated theories, too. Below we present a brief analysis of factors in ultrasonic wave propagation in soft tissues following Engelbrecht and Chivers (1989). For more details see the references therein.

It is perhaps a sign of the complexity of the problem that progress in the scientific sense in this field has been relatively slow. The initial models of tissue ultrasound interaction were based on (large-scale) planar interfaces between lossless media and have never been experimentally verified. The subsequent developments in electronic circuitry permitted a greater dynamic range of returned echoes to be visualized in so-called grey-scale scanners, the low level reflections corresponding to scattering from the small-scale inhomogeneities of the tissue volumes themselves (Chivers and Hill, 1975).

The extraction of quantitative information from the small-scale scattering signals has occupied considerable attention. There does not appear to be agreement yet as to the analytical formulation that should be used to obtain an objective parameter such as the backscattering cross-section per unit volume per unit solid angle. A number of different approaches have been suggested in the literature (Waag, 1984), but when a given set of measurement data is inserted into them, they give different results. The design of critical experiments in this area then becomes extremely difficult.

Recent developments have included analysis of reflections from media in which the acoustic properties vary continuously as a function of space, from interfaces of finite thickness, and from interfaces between media with different loss properties (attenuation). In addition there has been some interest in the distortion of the wavefronts propagating in soft tissue due to scattering mechanisms, and to the potential application of the concepts of coherent and incoherent scattering.

The scattering represents only one of the two contributors to the overall loss (or attenuation), the other being absorption by which ultrasonic energy is converted into the internal energy of the tissue at a molecular level. Knowledge of the attenuation is a prerequisite for making quantitative scattering measurements. For this reason, and because of its potential as a tissue characterization parameter in its own right, there has been considerable recent interest in the measurement of attenuation in vivo (Kuc, 1980). One of the major problems that arises (which is also inherent in the measurement of scattering), is that of the need to correct for the finite aperture of the transducer with some form of diffraction analysis.

The explicit assumption in almost all of the diagnostic analyses performed to date is that of linearity. The achievement of good resolution in the ultrasonic image obtained is dependent upon the production of very short pulses. The time average of these leads to low average intensity exposure levels, so that it has been believed that biological effects of thermal origin are not significant in diagnostic applications. Until relatively recently the pulses were considered to be too short (or too small) to simulate cavitation, but this view is under scrutiny with present evidence. In addition there has been a development of interest in the peak amplitude of the pulses produced by diagnostic machines which may lead to disruptive mechanical effects.

Muir and Carstensen (1980) appear to have been the first to identify the possibility of finite amplitude propagation in ultrasound used in biomedicine. Since then there has been a considerable increase in interest in the work governing theoretical and experimental investigation of focussed fields in water, the measurement of the non-linearity exhibited by diagnostic pulses in water, and the calculation of potential non-linear effects in tissues. For an effective calculation in the last of these areas, knowledge of the non-linear parameter for different tissues is required. The reader is referred to Bacon (1986), Bjørnø (1986), Humphrey et al. (1986) for details.

The complexity of ultrasonic wave propagation thus arises from a combination of factors. In the first place there is the biochemical sophistication of the media concerned, which is based on dynamic equilibria with the result that there may be a variety of manifestations possible to which we give the same label, e.g. liver, spleen, etc.

Second, there is the variety of physical and acoustic problems: the diffractive nature of the transducer fields, the presence of absorption, and of large-scale inhomogeneities and small-scale scatterers, and possible finite amplitude propagation effects. It has been the custom to deal with each of these problems individually, with a second problem being introduced as a perturbation of the results obtained for the first. While this has the attraction of simplicity, it is difficult to assess the overall effect of the different phenomena when present simultaneously. Engelbrecht and Chivers (1989) have tried to find a unified approach which permits many effects to be taken into account simultaneously. This Section is based on their results.

6.3.2. Statement of the problem

In ultrasonics the transducers are usually designed to generate or receive one type of wave. Mostly a longitudinal wave is generated but transverse waves are not excluded, although generated by a more sophisticated transmitter (Krautkramer and Krautkramer, 1977). Here we concentrate our attention on longitudinal waves only. In 1D approach, the scheme of transmitting and receiving a signal is shown in Fig. 6.7.

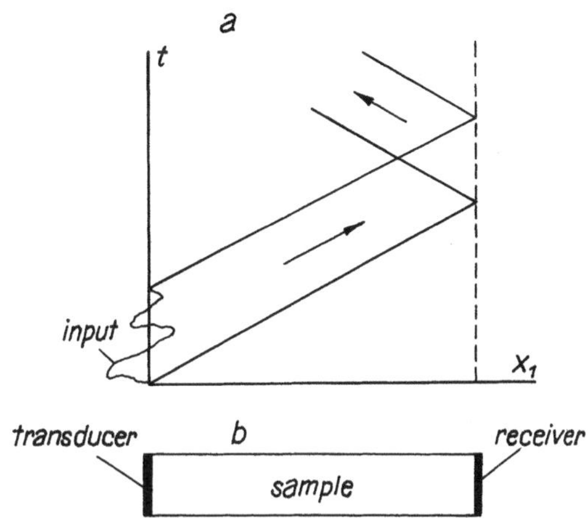

Fig. 6.7. Schematic diagram of 1D pulse propagation in the $X_1 - t$ plane (a) and the corresponding physical situation (b).

It is clear that such a scheme immediately evokes the attention to evolution equations (Section 4.2). Indeed, an evolution equation governs the propagation of just a single wave out of possible n waves in a properly chosen moving frame. It is easy to generalize the 1D approach to the 2D approach with or without circular symmetry, i.e. to 2D plane and cylindrical waves, generated by rectangular or circular transducers. However, for a usual piston-type transducer is well-known that two zones must be distinguished in a generated wave pattern: the near-field (Fresnel) and far-field (Fraunhofer) zones (see Krautkramer and Krautkramer, 1977). In the near field, the edge waves due to the transducer's boundary, and the plane wave generated from the surface of a transducer are combined in a complicated pattern, in the far field the influence of edge waves has largely died out. This is schematically shown in Fig 6.8, where Fig. 6.8a shows the fronts of edge and plane waves and Fig 6.8b demonstrates the pressure changes on the axis due to combined influence of edge and plane waves. We start here with the description of the plane wave (Section 6.3.3) and later explain the more complicated situation in the near field with both edge and plane waves.

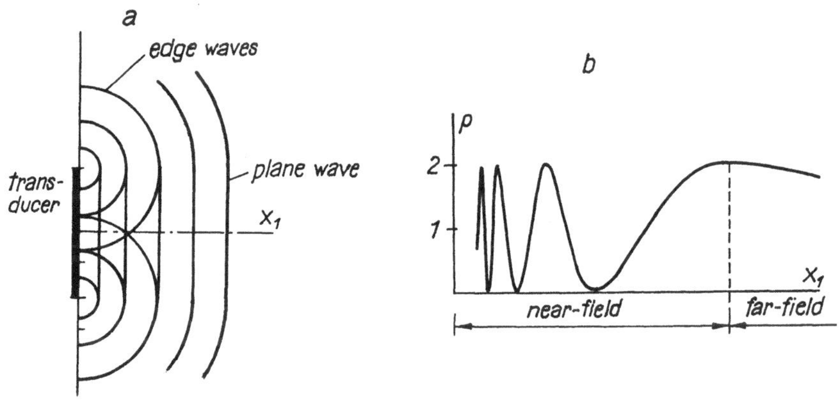

Fig. 6.8. Simplified scheme of edge and plane waves (a) and the pressure distribution on the axis X_1 (b) for a piston-type transducer.

The mathematical model in the form of a 2D evolution equation is derived according to the ideas presented in Chapters 2 and 4 using material coordinates. There is no difference in the general structure of the governing equations for fluids and solids when material coordinates are used, only the constitutive equation should be clearly stated. At this moment there is no need to discuss the structural behaviour of a soft tissue—a model will be derived on the basis of continuum mechanics involving non-linearity, inhomogeneity, attenuation, and relaxation.

Most of traditional models of soft tissues are based on the nonlinear constitutive equation of a fluid like

$$p' = A\frac{\rho'}{\rho_0} + \frac{B}{2}\left(\frac{\rho'}{\rho_0}\right)^2, \tag{6.56}$$

where p', ρ' are the changes of pressure and density, respectively, and ρ_0 is the initial density. The constants A and B are given (Bjørnø, 1986)

$$A = \gamma p_0 , \quad B = (\gamma - 1)c_0^2 p_0 , \tag{6.57}$$

where γ is the ratio of specific heats, and p_0 is the initial pressure. The ratio $B/A = k$ is usually called nonlinear parameter. In solids (see models of Chapters 2–4) we have $k = 1 + m_0$ where m_0 is the ratio of the third and second order elastic moduli.

Next, a soft tissue is a strongly inhomogeneous medium, with properties usually varying in certain intervals which are not clearly defined (Chivers, 1981). We shall further distinguish two classes of inhomogeneities:

(i) large scale inhomogeneity when the characteristic distance l is much greater than the wavelength τ_c;

(ii) small-scale inhomogeneity when the characteristic distance l is of the same order as wavelength τ_c. This situation is modelled in terms of a dispersive medium (see Section 5.3).

A special attention should be paid to the description of the absorption mechanism. It appears that the most general approach could be given in terms of a relaxing medium. This predicts an additional dispersion but the dissipation has a much more complicated character than the usual assumption of an exponential decay (see Section 5.1.3). It is not impossible that this exponential decay could be an additional mechanism added to some more general mechanism.

6.3.3. The mathematical model

There is no need to repeat here the considerations for deriving evolution equations as stated in Chapter 4. Referring also to Engelbrecht and Chivers (1989), we present here the final form of a 2D evolution equation meant for the description of a plane wave generated by an ultrasound transducer. In terms of scaled particle velocity v, the governing equation is

$$\frac{\partial}{\partial \xi} R[v(\xi, \tau)] = \Delta \left(\frac{\partial^2 v}{\partial \eta^2} + \frac{n}{\eta} \frac{\partial v}{\partial \eta} \right) , \tag{6.58}$$

where $n = 0, 1$ for planar and circular transducer, respectively. We denote by X_1 the coordinate along the sound beam and by X_2 the transverse coordinate. The moving frame (see Section 4.2) is then determined by

$$\tau = k v_0 X_1 / c_0 \tau_c , \tag{6.59}$$

$$\xi = (c_0 t - X_1)/\tau_c , \tag{6.60}$$

$$\eta = \varepsilon^{3/2} X_2 / r , \tag{6.61}$$

where k is the nonlinear parameter, v_0 and τ_c are the amplitude and the wavelength of the excitation, respectively, c_0 is the sound velocity, and r is effective radius of the transducer. The small parameter ε is related to the Mach number, i.e. $\varepsilon \sim k v_0 / c_0$.

The wave operator $R[v(\xi, \tau)]$ describes the propagation along the axis X_1, while the r.h.s. of Eq. (6.58) is responsible for the diffractional expansion in the transverse direction. The coefficient Δ measures the influence of the latter phenomenon, in solids it is given by the expression

$$\Delta = \frac{c_2^2}{c_0^2} \left(\frac{\tau_c}{r} \right)^2 , \tag{6.62}$$

where c_2 is the velocity of the transverse wave. The following form of the wave operator is proposed by Engelbrecht and Chivers (1989):

$$R(v) = \frac{\partial v}{\partial \tau} - v\frac{\partial v}{\partial \xi} + \Omega^{-2}\frac{\partial^3 v}{\partial \xi^3} - \theta(\tau)\frac{\partial v}{\partial \xi} + \Lambda v - \Gamma^{-1}\frac{\partial^2 v}{\partial \xi^2} - \Xi\frac{\partial}{\partial \xi}\int_0^{\xi}\frac{\partial v(z)}{\partial z}\exp\left(-\frac{\xi - z}{Z}\right)dz\,. \tag{6.63}$$

In this operator, nearly all the features identified in Section 6.3.2 are represented. The first term describes the linear wave propagation in a homogeneous lossless medium, while the second describes nonlinearity (see Chapter 2). The third term describes small-scale scattering, the parameter Ω being given by

$$\Omega^2 = 2k\frac{\tau_c^2}{a^2}\frac{v_0}{c_0}\,, \tag{6.64}$$

where a is the scatterer dimension. The fourth and fifth terms describe scattering from large-scale inhomogeneities for which the parameters are

$$\theta(\tau) = \frac{1}{2}g(\tau)\,, \tag{6.65}$$

where $g(X_1)$ is the function of the continuous changes of sound speed, and

$$\Lambda = \frac{1}{2}\tau_c\frac{\partial f}{\partial X_1}\,, \tag{6.66}$$

where $f(X_1)$ gives the continuous changes of the modulus of elasticity. The sixth term describes classical absorption, with

$$\Gamma = 2k\tau_c v_0/n_0\,, \tag{6.67}$$

where n_0 is the kinematic viscosity. A more complicated expression can be used here if thermal losses are also to be included. The seventh term covers relaxation effects for a particular relaxation time τ_0 with

$$\Xi = \frac{1}{2}\varepsilon_1/\varepsilon\,, \tag{6.68}$$

in which ε_1 indicates the difference between the instantaneous c_i and equilibrium c_0 velocities through $c_i^2 = (1+\varepsilon_1)c_0^2$, $\varepsilon_1 > 0$; and with

$$Z = \tau_0 c_0/\tau_c\,. \tag{6.69}$$

In the 1D approach the evolution equation is

$$R(v) = 0\,. \tag{6.70}$$

The tissue parameters (and functions) needed for describing the wave process according to Eqs. (6.58), (6.63) are as follows:

- equilibrium velocity c_0;
- density ρ_0;
- nonlinear parameter k (see above);
- small-scale parameter a;
- change in velocity $g(X_1)$;
- change in elasticity $f_1(X_1)$;
- kinematical viscosity n_0;
- instantaneous velocity c_i (or the parameter ε_1);
- relaxation time τ_0.

Equation (6.58) is required to be solved for a given initial condition

$$v(\xi,\tau,\eta)|_{\tau=T} = v(\xi,\eta) \tag{6.71}$$

that corresponds to the known distribution at certain $X_1 = L$.

There is no other way as to solve Eqs. (6.58), (6.63) numerically. However, in order to understand the possible effects, let us imagine a simplified wave operator

$$R(v) = \frac{\partial v}{\partial \tau} - \Xi \frac{\partial}{\partial \xi} \int_0^{\xi} \frac{\partial v(z)}{\partial z} \exp\left(-\frac{\xi - z}{Z}\right) dz \tag{6.72}$$

and the initial condition

$$v(\xi,\tau,\eta)|_{\tau=0} = \exp(-\eta^2)\sin\xi\,. \tag{6.73}$$

Then for small $Z_e \ll 1$ and $n = 1$ we obtain

$$v = \exp(-\Xi Z\tau)\exp(-m\eta^2)m^{1/2}\sin(\xi + \Xi Z^2\tau + \delta_1)\,, \tag{6.74}$$

$$m = (1 + 16\Delta^2\tau^2)^{-1}\,, \tag{6.75}$$

$$\delta_1 = -\eta^2 4m\Delta\tau + \arctan 4\Delta\tau\,. \tag{6.76}$$

Consequently, the solution has the following parts:
- attenuation due to dissipation ($\exp(-\Xi Z\tau)$);
- attenuation due to diffraction ($m^{1/2}$);
- increase in the beam width ($\exp(-m\eta^2)$);
- deformation of the surfaces of equal phase:
  - due to dispersion ($\Xi Z^2\tau$);
  - due to diffraction (δ_1).

It means that in case of a rectangular transducer, an initially plane wave is transformed into a cylindrical wave, and in case of a circular transducer, the resulting wave is a spherical one. Equations (6.58), (6.63) are possible models for ultrasound propagation in the far field. Below we also discuss briefly the peculiarities of the near field.

6.3.4. Complexity of the near field

The near field of an ultrasonic piston-type transducer has a complicated structure even in the linear approximation. Feldmann et al. (1990) have proposed to describe the near field as a sum of finite waves—a plane wave and an edge wave—each of these waves being described by its own evolution equation. The plane wave can be described by a 1D evolution equation because the diffraction effects are not important due to the short distance. The edge waves have, however, a very special structure, which can be explained by a detailed analysis of the Huygens-Rayleigh integral (Skudrzyk, 1971). This leads to the understanding that despite their geometrical similarity to cylindrical wavefronts, the behaviour of the edge waves is more complicated. The most important fact is that in the near field the amplitude of an edge wave does not decrease with distance from the source.

Indeed, the calculation of the linear sound field from a piston-type circular transducer is based on the Huygens-Rayleigh integral (Skudrzyk, 1971)

$$P = \frac{j\kappa\rho_0 c_0 v_0}{2\pi} \int_\sigma \frac{1}{r}\exp(-j\kappa r)d\sigma \tag{6.77}$$

where P is the sound pressure, κ is the wavenumber and r the distance from the field point under examination to an elementary area of the piston $d\sigma$.

As far as the velocity v is given by

$$v = \frac{j}{\kappa \rho_0 c_0} \frac{\partial P}{\partial X_1}, \tag{6.78}$$

then finally

$$v = v_0 \left[\exp(-j\kappa X_1) - \frac{X_1}{(X_1^2 + a^2)^{1/2}} \exp(-j\kappa (X_1^2 + a^2)^{1/2}) \right], \tag{6.79}$$

where a, as before, is the radius of the transducer. The first term in (6.79) represents the plane wave, the second—the edge wave. It is obvious that the increase in the amplitude A of the edge wave is proportional to

$$A = \frac{X_1}{(X_1^2 + a^2)^{1/2}} = \frac{X_1/a}{((X_1/a)^2 + 1)^{1/2}}. \tag{6.80}$$

This dependence is schematically shown in Fig. 6.9. Feldmann et al. (1990) proposed a piece-wise linear approximation of the function (6.80), composed by two straight lines (Fig. 6.9). For the part OK the evolution equation is proposed like

$$\frac{\partial u}{\partial \tau} + u \frac{\partial u}{\partial \xi} - \frac{u}{\tau} = 0, \tag{6.81}$$

for the part KL like

$$\frac{\partial u}{\partial \tau} + u \frac{\partial u}{\partial \xi} = 0, \tag{6.82}$$

where u is a scaled velocity. In terms of v, the solution of Eq. (6.81) is

$$v_e = v_0 \exp[j\kappa(c_0 t - (X_1^2 + a^2)^{1/2} - \tfrac{1}{2} v_e X_1]. \tag{6.83}$$

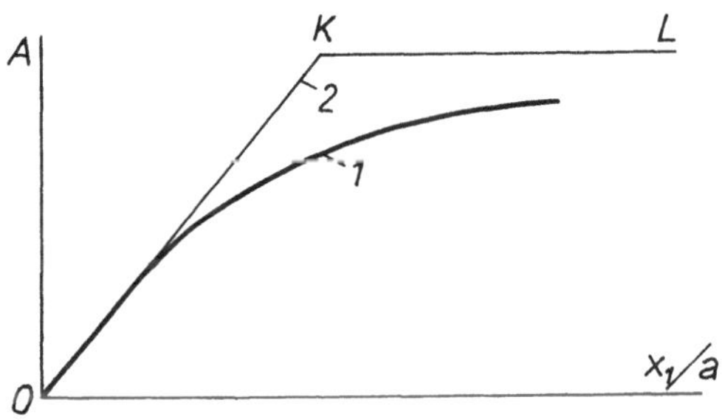

Fig. 6.9 Amplitude dependence on the distance: curve 1 – expression (6.80), curve 2 – approximated dependence.

From expression (6.79), near the transducer and multiplied by the time factor $\exp(j\omega t)$, the edge wave is governed by

$$v \sim v_0 \frac{X_1}{a} \exp[j\omega t - j\kappa(X_1^2 + a^2)^{1/2}]. \tag{6.84}$$

Both expressions (6.83) and (6.84) yield the similar result, as shown in Fig. 6.10 where the curves calculated by different expressions coincide with given accuracy. Matching the solutions at point K (see Fig. 6.9) is possible with a good accuracy, too, as shown by Feldmann et al. (1990).

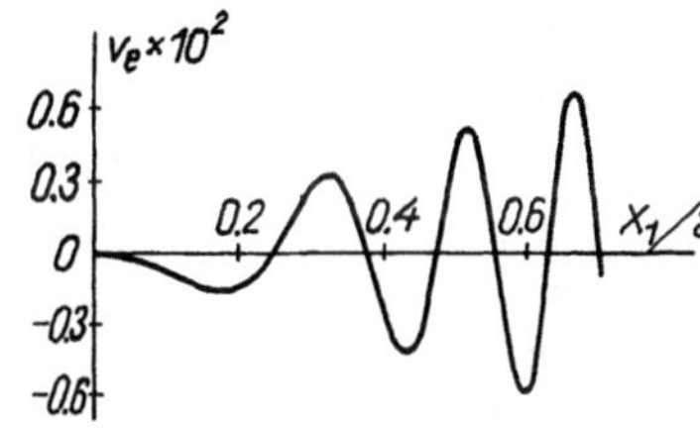

Fig. 6.10. Particle velocity v_e vrs. distance calculated from expressions (6.83) and (6.84); $v_0 = 0.01\text{ms}^{-1}$, $a = 0.01$ m, $\kappa = 8378\text{m}^{-1}$ (i.e. 2 MHz).

6.4. PROPAGATION OF NERVE PULSES

6.4.1. The physiological background

The pulse transmission in nerve fibres is a fascinating physical phenomenon that has already attracted researchers' attention for a long time. Isaac Newton has put the question about "...vibrations of this medium, excited in the brain..." (see Scott, 1975, p.487), being fairly close to the electrodynamics. Contemporary theory of nerve pulse transmission is based on the famous results of Hodgkin and Huxley (1952) who have explained the role of the ion currents responsible for the stable wave-profile. There are still many questions to be answered and the research in this field is developing fast, bearing also in mind the artificial neural networks.

The main structural feature in nerve pulse dynamics is a stable solitary pulse of a characteristical asymmetric profile (Fig. 6.11a) propagating at a speed of about 20 ms^{-1}. Solitary waves as carriers of energy are well understood in conservative media (see Section 5.3). Here the situation is different—the medium is highly dissipative but the source terms balance the energy outflux resulting in a solitary wave. The physiological background is the following (see also Fig. 6.11b). The nerve pulse (voltage) is transmitted down the axoplasm core of a nerve which is surrounded by a cylindrical membrane. The relative concentration of ions (mainly sodium and potassium) create the transmembrane potential. If an electric stimulus is applied to the nerve then the membrane acts in different ways depending on the value of the stimulus. If the stimulus is below a certain threshold value, then the depolarisation process of the membrane is reversible and the equilibrium state returns fast without any pulse propagating. If, however, the stimulus is above this threshold, the inward flow of the sodium ions starts. This process is followed then by an increase in the potassium permeability which causes an outward flow of the potassium ions. Later due to the balance of inward and outward flows the process returns to the equilibrium again but through an overshoot. This all results in an asymmetric solitary wave

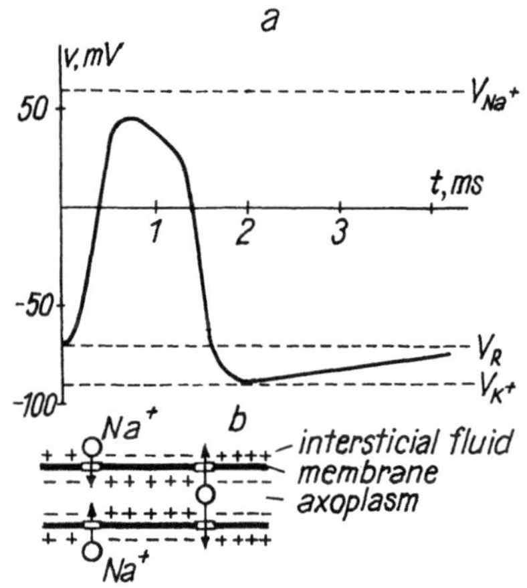

Fig. 6.11. A schematical nerve pulse: a – a typical profile, b – idealised nerve with ion currents. Dashed lines – values of resting potentials.

shown in Fig. 6.11 and propagating without changes in its profile, i.e. representing a constant profile solution.

The main point in wave dynamics as explained above is the concept of finite velocities—every excitation propagates with a finite velocity determined mainly by the properties of the medium. This physical understanding is reflected in the hyperbolicity of the governing equations in wave dynamics.

Nerve pulse is without any doubt a wave. However, an interesting paradox appears here—the classical theory of nerve pulse transmission is based on the parabolic equations which strictly speaking lead to infinite velocities. Certainly, the situation is more complicated while the governing equations consist of a nonlinear source term responsible for modelling ion currents. It is shown then that in this case, the governing equations have a progressive type solution for which the finite velocity exists (Cole, 1972; Scott, 1975). In this way the paradox is seemingly solved but only theoretically. The constant profile solution corresponds to a separatrix in the phase portrait and, in order to follow it, a physically abnormal accuracy of calculations is needed. In other words, this accuracy is needed for determining the wave velocity that should be given according to Cole (1972) with an accuracy of 10^{-18}. Even in this case, some special methods are needed in order to calculate a full profile (Miura, 1982). This situation leads back to inspecting the governing equations in order to estimate the validities of basic assumptions and to find other possibilities to represent the main features of the process. Such an approach is also quite logical in the general sense because once a certain theory is settled, it is used widely without questioning until some of its shortcomings become vital. Then, taking into account the progress in physical understandings, mathematical methods, etc., the next step can be made, at least to get rid of some shortcomings.

Before presenting some relatively novel ideas in modelling the nerve pulse transmission it must be stressed that nerve pulse is usually classified as a dissipative structure (Scott, 1975), modelled by the parabolic (diffusion-type) system with source terms. These models are widely used in biology and chemistry (Othmer, 1986) and the notion of dissipative structures itself belongs to Prigogine (see Prigogine and Stengers, 1984). On the other hand, solitonic structures belong to conservative systems (see Section 5.3). There is an urgent need to make very clear distinction between the corresponding systems as well as between the solitary waves the origin of which is not always clear from physical observations.

The first idea is to reinspect the basic mathematical models governing the nerve pulse transmission. The usual approach is to start from telegraph equations and neglect the inductance. This means that the hyperbolicity is neglected from the very beginning and the natural result is the parabolic equation which in terms of solid mechanics corresponds to the heat conduction equations with an additional source term. It must be stressed that the conventional theory of nerve pulse transmission was established in the fifties of this century when the knowledge about solitary waves was limited. Note that the concept of solitons was introduced only in 1965 by Zabusky and Kruskal and later the intensive research began including the development of reductive methods in order to construct one-wave equations, i.e. evolution equations (Taniuti and Nishihara, 1983). The evolution equations, as a rule, are given in a moving frame that stems from the hyperbolicity of the basic governing system. Quite naturally, the elegant approaches from the soliton theory were not used for parabolic equations derived for nerve pulse dynamics. Lieberstein (1967) has used the natural form of telegraph equations in nerve pulse dynamics solving the problem numerically. In the eighties the time was ripe to apply the ideas of solitonic structures in neighbouring areas. The evolution equation for a nerve pulse was derived by Engelbrecht (1981) and its various aspects analysed later (see summary by Engelbrecht, 1991). As a result, a novel second-order evolution equation typical for nerve pulse transmission gives a natural prolongation to the family of known but mainly first-order evolution equation.

The quest for possible unification of dissipative and solitonic structures has recently got an important milestone in the framework of continuum mechanics (Maugin, 1990). It has been shown that starting from the Lagrangian formulation and introducing a dissipation potential, it is possible from a common starting point to derive either diffusion-type systems that may exhibit dissipative structures or conservative systems (perturbed by dissipation) which may exhibit solitonic structures.

Here we shall present the conventional mathematical models with a brief analysis and then follow the idea of evolution equations in order to derive the nerve pulse evolution equation. Following Maugin and Engelbrecht (1994), we use the formalism of internal variables (see Section 2.4) for deriving the governing equations. Some numerical examples given demonstrate the richness of the simple evolution equation governing the nerve pulse transmission.

6.4.2. Conventional mathematical models

The conventional initial system of an axon is the following (Scott, 1975)

$$\pi a^2 C_a \frac{\partial v}{\partial t} + \frac{\partial i_a}{\partial x} + 2\pi a I = 0 \,, \tag{6.85}$$

$$\frac{\partial v}{\partial x} + \frac{R}{\pi a^2} i_a = 0 \,, \tag{6.86}$$

where the notations of Lieberstein (1967) are used: v is the potential difference across the membrane, i_a is the axon current per unit length and a is the axon radius. Further, C_a is the axon self-capacitance per unit area per unit length, R is the specific resistance and I is the ion current density. It is easy to derive a second-order equation in terms of v

$$\frac{\partial^2 v}{\partial x^2} = RC_a \frac{\partial v}{\partial t} + \frac{2}{a} RI \,, \tag{6.87}$$

which can easily be compared with the heat conduction equation.

The most ingenious result by Hodgkin and Huxley (1952) was to give an empirical expression for the ion current $j_i = 2\pi a I$ depending on (as they called them) phenomenological variables n, m, h. These variables govern: n – the potassium conductance (turning on), and m, h – the sodium conductance (turning on and turning off, respectively). The ion current expression is (Hodgkin and Huxley, 1952):

$$j_i = g_K n^4 (v - V_K) + g_{Na} m^3 h (v - V_{Na}) + g_L (v - V_L) \,, \tag{6.88}$$

where g_K, g_{Na} and g_L are potassium, sodium and leakage conductances, respectively, and V_K, V_{Na} and V_L are the corresponding equilibrium potentials. The variables n, m and h are governed by the following kinetic equations:

$$\dot{n} = -(n - n_0(v))/\tau_n(v) \,, \tag{6.89}$$

$$\dot{m} = -(m - m_0(v))/\tau_m(v) \,, \tag{6.90}$$

$$\dot{h} = -(h - h_0(v))/\tau_h(v) \,. \tag{6.91}$$

Here n_0, m_0 and h_0 are equilibrium values, and τ_n, τ_m and τ_h are characteristics relaxation times. As a result of careful experimentation, the empirical relations for these parameters are found. From this time on two main directions in the theory of nerve pulse transmission can be distinguished:

(i) to make the mathematical model even more complicated in order to account for all the possible physical and physiological effects;

(ii) to make the mathematical model simpler in order to get simple solutions still preserving the main physical features of a pulse.

Direction (i) is more developed in physiology and cardiac dynamics where the influence of various drugs ("narcotizing") on nerve systems is of interest. The DiFrancesco-Noble (1985) model incorporates as many as twelve different terms in the total ion current, accounting for delayed potassium current, background potassium and sodium currents, background calcium current, potassium-sodium exchange current, etc. The number of phenomenological variables may certainly be even higher.

Direction (ii) has got its best representation in the Fitz-Hugh-Nagumo (see Nagumo et al., 1962) model. This model accounts only for one additional variable and the ion current is taken in the form

$$j_i = k_1 v + k_3 v^3 + w\,, \tag{6.92}$$

where k_1 and k_3 are constants. The variable w is usually referred to as a recovery variable and it is governed by a kinetic equation of the following type:

$$\dot{w} + c_0 \gamma w = q_0 (v + q_1)\,, \tag{6.93}$$

where c_0, γ, q_0 and q_1 are constants. Nagumo et al. (1962) have used $\gamma = 0$ resulting in a final equation

$$\frac{\partial^4 v}{\partial t \partial x^3} = \frac{\partial^2 v}{\partial t^2} + \mu(1 - v + \varepsilon v^2)\frac{\partial v}{\partial t} + v\,, \tag{6.94}$$

where μ and ε are again constants. Later the form with $\gamma \neq 0$ was found to be preferable but still many papers are devoted to the analysis of equation (6.94) and its variants. There have also been attempts to replace the cubic polynomial with respect to v in relation (6.92) by a piece-wise linear approximation in order to get simpler mathematical expressions (McKean, 1970).

6.4.3. Evolution equation for nerve pulse transmission

As said in Section 6.4.1, the conventional approach has several shortcomings. The point is that the governing equation is of the parabolic type (Section 6.4.2). Therefore one could act just in a simple way putting the neglected terms back. This was done by Lieberstein (1967) who used the hyperbolic telegraph equations for numerical calculations. Engelbrecht (1981) has derived an evolution equation for the nerve pulse starting also by the hyperbolic telegraph equations. This formalism is given in Section 4.2 in detail. Here we present only the brief analysis in order to demonstrate the idea of evolution equations as applied to nerve pulses. The reader is referred to Engelbrecht (1991) for more details.

Putting the neglected terms back means simply that instead of Eqs. (6.85), (6.86) one has to use

$$\pi a^2 C_a \frac{\partial v}{\partial t} + \frac{\partial i_a}{\partial x} + 2\pi a I = 0\,, \tag{6.95}$$

$$\frac{L}{\pi a^2}\frac{\partial i_a}{\partial t} + \frac{\partial v}{\partial x} + \frac{R}{\pi a^2} i_a = 0\,, \tag{6.96}$$

where L is the axon specific self-inductance. The second-order equivalent to Eqs. (6.95), (6.96) is

$$\frac{\partial^2 v}{\partial x^2} - LC_a \frac{\partial^2 v}{\partial t^2} = RC_a \frac{\partial v}{\partial t} + \frac{2}{a} RI + \frac{2}{a} L \frac{\partial I}{\partial t}\,. \tag{6.97}$$

The l.h.s. of Eq. (6.97) shows that the velocity should be

$$c_0 = (1/LC_a)^{1/2}\,. \tag{6.98}$$

However, due to the r.h.s. of Eq. (6.97) the real velocity must be different from that. According to the ideas presented in Section 4.2, we still can use velocity c_0 to build up the moving frame. By doing so and introducing the assumptions of slow changes (see Engelbrecht, 1991), we obtain for $v = v(x, \xi)$, $\xi = c_0 t - x$ from the "two-wave" system (6.95), (6.96) a "single-wave" evolution equation

$$c_0 \frac{\partial v}{\partial x} + \frac{1}{2} m_4 v + \frac{1}{2} m_2 I = 0\,, \tag{6.99}$$

where m_2 and m_4 are constants. Now, for ion current I the usual models are used. According to Hodgkin and Huxley (1952), we should take three additional phenomenological variables into account (Eq. (6.88)) but following Nagumo et al. (1962) only one additional variable is sufficient losing certainly the accuracy.

Together with the ion current (6.92) and "constitutive equation" (6.93) for variable w (with $\gamma = 0$) the resulting equation in suitably chosen variables reads

$$\frac{\partial^2 z}{\partial \xi \partial x} + f(z) \frac{\partial z}{\partial \xi} + g(z) = 0\,, \tag{6.100}$$

$$f(z) = b_0 + b_1 z + b_2 z^2\,, \tag{6.101}$$

$$g(z) = b_{00} z\,, \tag{6.102}$$

where b_0, b_1, b_2 and b_{00} are constants.

This is the sought simplest second-order evolution equation for a nerve pulse transmission that must be solved under initial excitation $z(0, \xi)$ and the proper boundary conditions. The physics of the nerve pulse predicts the following properties for the quadratic polynomial $f(z)$: the roots of $f(z) = 0$ are

$$z_1 > 0\,, \quad z_2 > 0\,, \quad z_1 \neq z_2\,. \tag{6.103}$$

Function $g(z)$ is usually linear but the modified models may involve a cubic polynomial $g(z)$ (Engelbrecht, 1991).

Equation (6.100) with expressions (6.101), (6.102) is able to describe:

- the solitary waves of asymmetric character;
- the all-or-none phenomenon, i.e. the existence of a threshold;
- refractoriness, i.e. a second action potential cannot be generated if the second stimulus is applied too soon after the first one.

One should notice that Eq. (6.100) by its nature is a "single-wave" equation and as the nerve pulses propagate all with the same speed, it can describe neither the overtaking which is physically inadmissible nor annihilation because it does not describe the head-on collision.

Two remarks are appropriate here. First, for a full kinetic equation (6.93) an equation with wave hierarchy (Whitham, 1974) follows:

$$\frac{\partial^2 z}{\partial \xi \partial x} + f(z) \frac{\partial z}{\partial \xi} + g(z) + \gamma \left(\frac{\partial z}{\partial x} + F(z) \right) = 0\,, \tag{6.104}$$

where $F(z) = \int f(z) dz$.

Second, the stationary wave in terms of $\eta = x + \theta \xi$, $\theta =$ const. is described by an ordinary differential equation of the Liènard type

$$z'' + f(z) z' + \theta^{-1} g(z) = 0\,, \tag{6.105}$$

where $(\;)' = \partial / \partial \eta$. The conditions (6.103) lead to a pulse-type solution and the origin in the phase plane is a stable node. The pulse-type solution is obtained with $z(0) = 0$, $z'(0) \neq 0$ that corresponds to a 2D projection from the 3D phase space of the original equation. This projection also explains the need for $z'(0) \neq 0$. It is easily understood that the celebrated van der Pol equation is of the same (6.105) type, only the conditions

$$z_1 < 0\,, \quad z_2 > 0 \tag{6.106}$$

must be obeyed. Consequently, the origin is an unstable focus but there exists a stable finite cycle. The full map of the Liénard equation (6.105) contains both cases (Engelbrecht, 1991).

In some sense the situation can be compared to the case of a linear oscillator with weak and strong damping. In the case of weak damping the motion is still oscillatory (c.f. with the limit cycle) and in the case of strong damping there is no oscillatory character at all (c.f. with pulse-type solution of (6.105)).

Numerical simulation of Eq. (6.100) and its stationary counterpart (6.105) demonstrates the properties described above. As examples, the formation of transient pulses is shown in Fig. 6.12. Figure 6.12a shows the formation of a pulse when the initial excitation has an amplitude above the threshold but below the asymptotic value. Fig. 6.12b shows the same but the initial excitation is above the asymptotic value. The outcome in both cases is the pulse with the same asymptotic value (c.f. Nagumo et al., 1962). Figure 6.13 shows the stationary profile and the corresponding phase plane for two cases: slightly below the threshold and slightly above the threshold. As expected, the outcome is drastically different. For more details the reader is referred to Engelbrecht (1991).

6.4.4. Internal variables in nerve pulse dynamics

Let us return to the notions of observable and internal variables (see Section 2.4). In nerve pulse dynamics, as a rule, the potential (voltage) is the observable variable. In addition to that, the usual notation involves either phenomenological (Hodgkin and Huxley, 1952), recovery (Nagumo et al., 1962), or else, auxiliary (Schierwagen, 1991) variables. These phenomenological, etc. variables describe the turning off and on ion conductances. In the classical Hodgkin-Huxley model the number of these variables is three (see Section 6.4.2); the modern theories include more variables taking the influence of background and exchange currents, etc. (DiFrancesco and Noble, 1985) into account. In the simplified FrizHugh-Nagumo model, however, there is only one "recovery" variable (see Section 6.4.2).

In terms of thermodynamics, these phenomenological or recovery or auxiliary variables are indeed internal variables which are not directly accessible to observation. It is of interest whether the formalism of internal variables (Maugin, 1990) can be used to clarify mathematical models of nerve pulse transmission from a thermodynamical viewpoint. Below we follow Maugin and Engelbrecht (1994) and present the general ideas of such an analysis.

We take the voltage v as an observable variable and consider the recovery variable w as an internal variable. The observable variable should be governed by a hyperbolic model and we expect (with normalized units) a field equation of the following type

$$\mu v_{tt} - v_{xx} = \mathcal{F}(v, v_t, w)\,, \tag{6.107}$$

$$w_t = R(v, w) \tag{6.108}$$

with $\mu > 0$, but possibly very small.

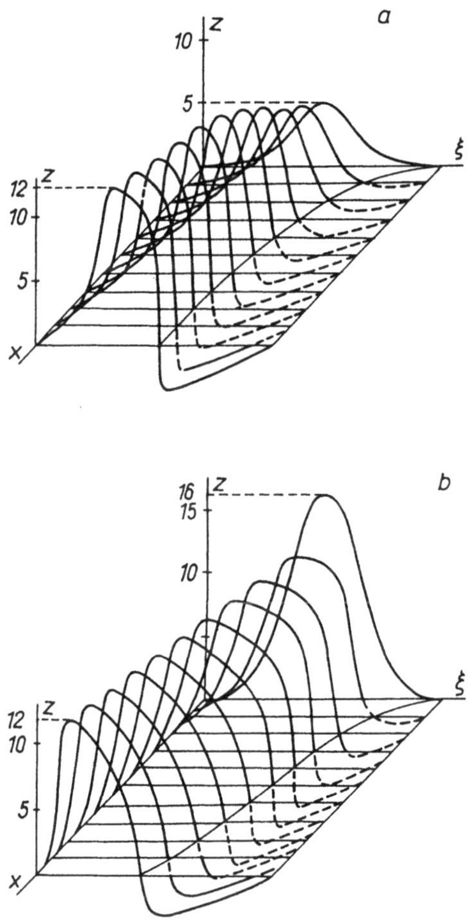

Fig. 6.12. The formation of transient pulses with initial excitation above the threshold (a) and above the asymptotic value (b).

The first of these reminds us of some type of generalized Frenkel-Kontorova model (Kosevich, 1988) for a linear elastic chain acted upon by a force $\mathcal{F}$ due to a substratum (interaction with the surroundings) but instead of a simple $\mathcal{F}(v) = \sin v$, we have a more complex force expression involving both friction (through v_t) and interaction with an additional field (w), the latter being governed by the relaxation equation (6.108). We naturally introduce a Lagrangian density $\mathcal{L}$ such as

$$\mathcal{L} = \frac{1}{2}\mu v_t^2 - W(v, v_x, w) = \frac{1}{2}\mu v_t^2 - \frac{1}{2}v_x^2 - \bar{W}(v, w), \tag{6.109}$$

in which we recognize kinetic, elastic and interaction energies. Contrary to Section 2.4 we do not grant any diffusional property to w so that W does not involve w_x. Nor is any inertia associated with w (internal variable). In the absence of dissipation

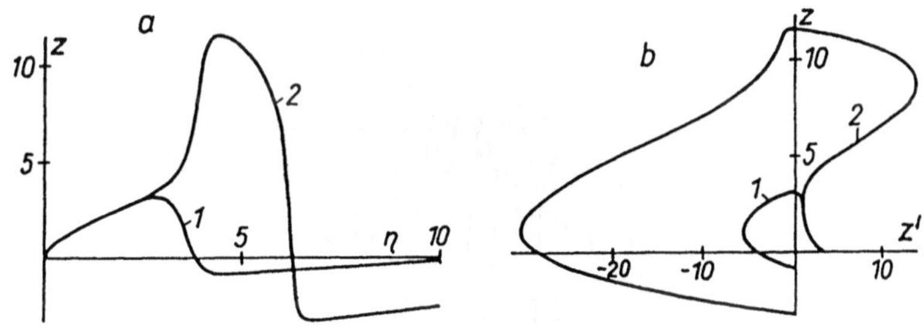

Fig. 6.13. Stationary profiles (a) and the corresponding phase plane (b); initial excitation slightly below the threshold (curve 1) and slightly above the threshold (curve 2).

and discarding thermal effects, Gibbs' equation reads

$$-\frac{dW}{dt} - \mathcal{F} v_t + \sigma v_{xt} = 0\,. \tag{6.110}$$

This is valid for any v_t and $\partial(v_x)/\partial t$ so that with the special form of W indicated in the second of equations (6.109), we get the so-called laws of state

$$\mathcal{F} = \bar{\mathcal{F}} \equiv -\partial \bar{W}/\partial v\,, \tag{6.111}$$

$$\sigma = \partial W/\partial v_x = v_x\,, \tag{6.112}$$

while the Euler-Lagrange equation of motion takes on the form (6.107) but with $\mathcal{F}$ reduced to $\bar{\mathcal{F}}(v)$. In the presence of dissipation, i.e. when (6.110) is replaced by the Clausius-Duhem inequality

$$-\frac{dW}{dt} - \mathcal{F} v_t + \sigma v_{xt} \geq 0\,, \tag{6.113}$$

we shall assume that there are neither "viscous" stresses nor "anelastic" strains, so that (6.112) continues to hold. Accounting for the general expression of $\bar{W}$, (6.112) and the fact that $\mathcal{F}$ in general depends on v_t, from (6.113) we deduce the following dissipation inequality

$$\phi = -\mathcal{F}^d v_t + \mathcal{R} w_t \geq 0\,, \tag{6.114}$$

where we have defined $\mathcal{F}^d$ and $\mathcal{R}$ by

$$\mathcal{F}^d = \mathcal{F} + \partial \bar{W}/\partial v\,, \quad \mathcal{R} = -\partial \bar{W}/\partial w\,. \tag{6.115}$$

Now we use the simplest theory of irreversible processes to formulate laws which govern $\mathcal{F}^d$ and $\mathcal{R}$ in agreement with (6.114). Thermodynamical equilibrium (Eq. (6.110)) corresponds to

$$\mathcal{F}^d = 0\,, \quad \mathcal{R} = 0\,. \tag{6.116}$$

Outside (but only slightly outside) equilibrium we assume the existence of a positive dissipation potential $\mathcal{D}$, which is homogeneous of degree two in v_t and w_t and such that

$$\mathcal{F}^d = -\partial \mathcal{D}/\partial v_t\,, \quad \mathcal{R} = \partial \mathcal{D}/\partial w_t\,. \tag{6.117}$$

Inequality (6.113) is automatically satisfied while the field equation for v and the evolution equation for w indeed take on the form

$$\frac{\delta \mathcal{L}}{\delta v} - \frac{\partial \mathcal{D}}{\partial v_t} = 0 , \quad \frac{\partial \bar{W}}{\partial w} + \frac{\partial \mathcal{D}}{\partial w_t} = 0 , \tag{6.118}$$

in perfect agreement with the general scheme presented at the Section 2.4. Accounting for the homogeneity of the dissipation function, it may be more convenient to introduce a dual potential $\mathcal{D}^*$ via a partial Legendre-Fenchel transformation such as

$$\mathcal{D}^*(v_t, \mathcal{R}) = \sup_{w_t} [\mathcal{R} w_t - \mathcal{D}(v_t, w_t)] \tag{6.119}$$

with

$$\mathcal{F}^d = \partial \mathcal{D} / \partial v_t , \quad w_t = \partial \mathcal{D}^* / \partial \mathcal{R} . \tag{6.120}$$

By way of example, consider the following dissipation potential $\mathcal{D}^*$ and interaction energy $\bar{W}$:

$$\mathcal{D}^* = -\frac{1}{2}\alpha v_t^2 + \frac{1}{2}\beta \mathcal{R}^2 , \quad \alpha \geq 0 , \quad \beta \geq 0 , \tag{6.121}$$

$$\bar{W}(v, w) = \mathcal{W}(v) - vw + \frac{1}{2}\gamma w^2 , \tag{6.122}$$

whence

$$\mathcal{F}^d = -\alpha v_t , \quad w_t = \beta \mathcal{R} . \tag{6.123}$$

Setting then

$$f(v) \equiv -\partial \mathcal{W} / \partial v , \quad \tau_w \equiv (\beta\gamma)^{-1} , \quad \omega_0(v) \equiv \gamma^{-1} v , \tag{6.124}$$

we obtain equations (6.107), (6.108) in the suggestive form

$$\mu v_{tt} - v_{xx} + \alpha v_t = f(v) + w , \tag{6.125}$$

$$w_t = -\frac{1}{\tau_w}[w - w_0(v)] , \tag{6.126}$$

which are the two equations of the Fitz-Hugh-Nagumo model (although the standard FHN-model has a minus sign in front of w in (6.125), but this can be made to agree by a change in the polarity of the voltage v). If w indeed is an internal variable, then the fact that we remain close to equilibrium during true thermodynamical excursions requires that this internal variable relaxes fast enough compared to a macroscopic time scale so that the thermodynamical evolution may be viewed as a sequence of constrained equilibria (Maugin, 1990). In more precise terms, if τ_w is the typical relaxation time of w, i.e., $\tau_w \cong | w | / | w_t |$, while $\tau_v = | v | / | v_t |$, is the characteristic time related to the observable variable or the external loading process, then the above theoretical construct is meaningful, on a thermodynamical basis, only in so far as the Deborah number $\mathcal{D}e$ defined by $\mathcal{D}e = \tau_w / \tau_v$ is very small, i.e.,

$$\mathcal{D}e \ll 1 . \tag{6.127}$$

This has further consequences. For instance, comparing equations (6.125), (6.126) and an appropriate rewriting (6.97) as

$$\mu v_{tt} - v_{xx} + \alpha v_t = f(v) + w + \lambda w_t , \tag{6.128}$$

we see that the generalization (6.128) of (6.125), (6.126) reduces to (6.125) on account of (6.127) on the simple condition that $\alpha / \lambda = O(| w | / | v |)$. Along similar lines, setting both λ and μ equal to zero corresponds to vanishing inductance as indicated by equation (6.97).

A natural special case of the above modelling corresponds to $\gamma = 0$ (no self-energy related to w) in which case (6.126) is replaced by the equation (not a relaxation equation per se)

$$w_t = \beta v \,. \tag{6.129}$$

This is an often considered simplified model (Nagumo et al., 1962).

Engelbrecht and Maugin (1994) have also analysed other models, demonstra-ting so the applicability of the formalism of internal variables to nerve pulse dynamics. This makes a basis for building up thermodynamically admissible models.

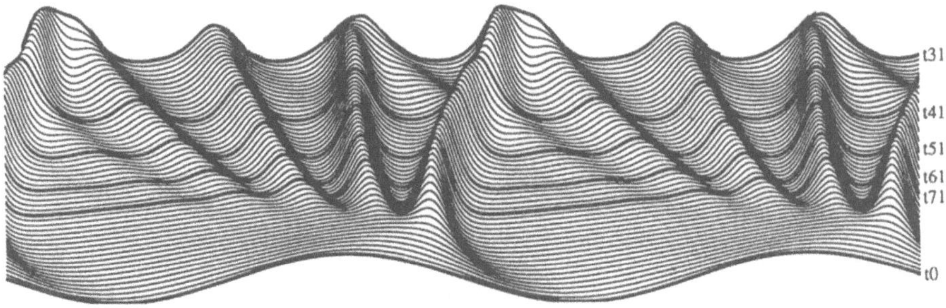

7. Essays: what is all that about

7.1. NONLINEARITY—PROPERTY OR NONPROPERTY

Linearity means that the rule that determines what a piece of a system is going to do next is not influenced by what it is doing now. Nonlinearity confers a certain flexibility upon the adaptability of an object to desirable behaviour.

M.Feigenbaum

The usage of suffices reflects the way how the mankind has formed the notions. The suffix "non" means negation and that makes an interesting philosophical "first move". Something was known and primary, and then the secondary notion was introduced that was different from the primary. Several questions arise immediately like why this secondary notion is needed, is the primary still basic or has the secondary overshadowing the primary and has it now taken the first (basic) place in our understandings.

These questions are really interesting because contemporary physics is in many senses the physics of "nons". A long list of notions could be compiled, all with the suffix "non": starting from non-Euclidean and non-Riemannian and coming to nonmonotonicity, nonuniqueness, nonlocality, nonlinearity, nonsingularity, etc. This is not the only possibility to emphasize negation. Another list could be easily added to the first, involving anisotropy, asymmetricity, infiniteness, irreversibility, instability, etc. Both lists are almost unlimited! We may call these notions "nonproperties" and ask a question: are nonproperties bad or good, or are they sometimes just nonsense? In more general sense of "non" we should turn to Bachelard (1934, 1940). He has shown how the idea of improving our understandings has moved from the philosophy of "why not" (Bachelard, 1934) to the philosophy of "non" (Bachelard, 1940). In physics, this is the usual way. Said Bachelard (1934):

"We shall always point out the dilemmatic character of new theories, like non-Euclidean geometry, non-Archimedean mass, non-Maxwellian physics (with Bohr), and arithmetics of noncommutative operations that can be called non-Pythagorean."

In this essay I shall concentrate on one of the nonproperties which has an enormously important place in contemporary physics—that is nonlinearity. The importance of being nonlinear, accepted during the last decades of the 20th century has profoundly changed physical sciences because of the qualitative richness and deepness of "primary", i.e. linear processes.

G.A.Maugin (1992b) has analysed nonlinearity from the viewpoint of nonscientists (again non!). Sharing his views, I would like to enlarge some of those and add a few of my own.

What is nonlinear? Maugin (1992b) has answered simply—nonlinear is what is not linear. This seems to be the natural way to deal with nonproperties but there is then the need to define the "primary" notion, i.e. linearity in our case.

Linearity means first of all proportionality—input and output (whatever the system is) are proportional. If input a has an output $f(a)$ and input b, respectively, $f(b)$, then the input $a+b$ gives the output $f(a)+f(b)$. Or in other words, every linear problem can be reduced to its single bricks. Then we solve the single (possibly simple) problems and then sum up the results. Only, said A.Toffler (1984): "Dissection is one of the most highly developed skills in contemporary Western civilization: the split-up of problems into their smallest possible components. We are good at it. So good, we often forget to put the pieces back together again."

All this does not work when nonlinearity is involved. I shall demonstrate what happens, on a simple example using a scheme action—response. With one dependent variable we may write $y_i = f_i(x)$, etc. The linear response is schematically illustrated in Fig. 7.1a. The larger the action, the larger the response and vice versa. If nonlinearity is weak, then a typical response curve is just inclined from the linear case (Fig. 7.1b). This is typical for nonlinear elastodynamics (Bell, 1973). Despite the weakness, the effect of nonlinearity is essential, leading to shock waves (Bland, 1969). "True" nonlinearity may induce more changes in the response. In Fig. 7.2 two cases are depicted: Fig. 7.2a shows how one response could be affiliated to several actions, Fig. 7.2b—how one action could cause several responses, some stable, some unstable. This is sufficient for the present analysis while more about such a nonlinear phenomenon, called hysteresis, can be found elsewhere (see, for example, Maugin, 1994a).

Time is a variable in all the processes around us. The influence of nonlinearity in the cause of time is even more remarkable than in examples described above. In dynamics, variables depend on time like $x(t), y(t), z(t), \ldots$ and processes, as a rule, are governed by differential equations. One of the tools in dynamics is the phase-space analysis that means geometrical interpretation of a process in $\{x, y, z, \ldots\}$ space. Every process is then depicted by a line (trajectory) in this space starting from a point where $t = 0$ and proceeding until given time. In linear world, these lines either proceed to one equilibrium point, or form a closed circle, or just diverge to infinity. In nonlinear world, there are not only more possibilities, but the trajectories also behave qualitatively differently. There may be many equilibrium points, the

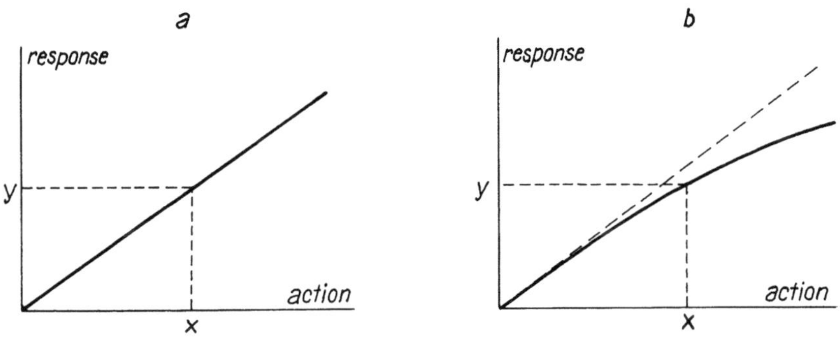

Fig. 7.1. Linear (a) and weakly nonlinear (b) responses.

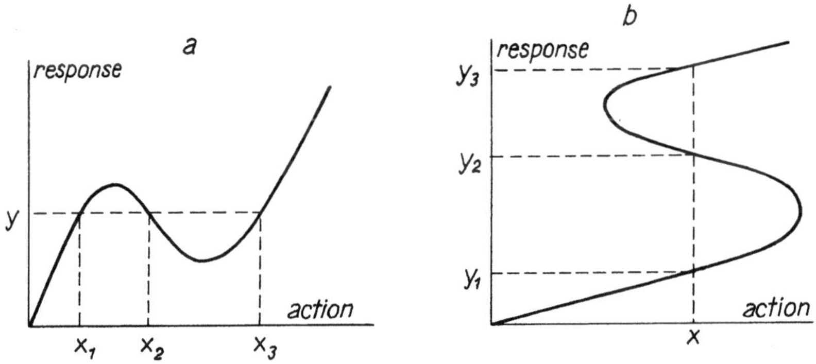

Fig. 7.2. Nonlinear responses.

trajectories may form complicated patterns, and the dependence on initial conditions may be decisively important. Some schematic examples are shown in Figs. 7.3–7.5. Fig. 7.3a shows the trajectories in $\{x, y\}$ plane with three qualitatively different outcomes: depending on the initial conditions y_1, y_2, y_3, the process either tends to $y = 0$, or $y = y_A$, or $y \to \infty$, respectively. In Fig. 7.3b, a three-dimensional phase space $\{x, y, z\}$ is shown with two trajectories. Trajectory I behaves monotonously, trajectory II has a qualitatively different path, all because the starting point is different.

Let now a nonlinear system depend also upon a parameter q that may take different values. Typical schematic responses $x(t)$ for two parameters q_1 and q_2 are shown in Fig. 7.4. The system may lead to the regular behaviour (Fig. 7.4a for q_1) and to the chaotic behaviour (Fig. 7.4b for q_2) where the predictability is lost. The phase diagrams in $\{x, \dot{x}\}$ plane are shown in Fig. 7.5a,b for the same cases, respectively. The trajectories, shown in Fig. 7.5b form a "strange attractor" that is typical of chaotic systems and can never happen in linear world.

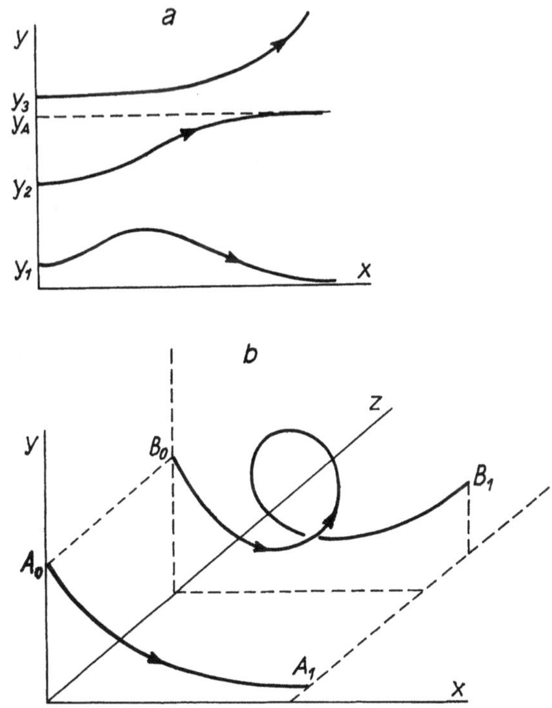

Fig. 7.3. Phase diagrams: a – three equilibrium points, b – different behaviour for different starting points A_0 and B_0.

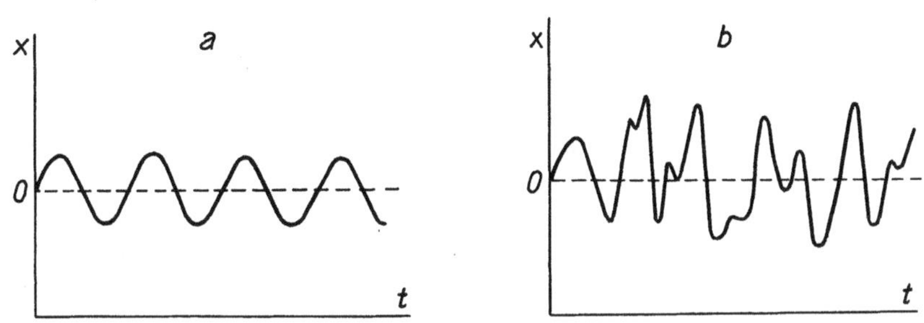

Fig. 7.4. $x(t)$ diagrams: a – regular, b – chaotic.

Many well-known examples like Lorenz equations, Rössler equations, driven Duffing equation, etc. behave in a way described above (Thompson and Stewart, 1986).

If the world is so nonlinear, why then the mankind started from the linear views? It is actually quite clear when looking back in time. Leonardo da Vinci (1452–1519)

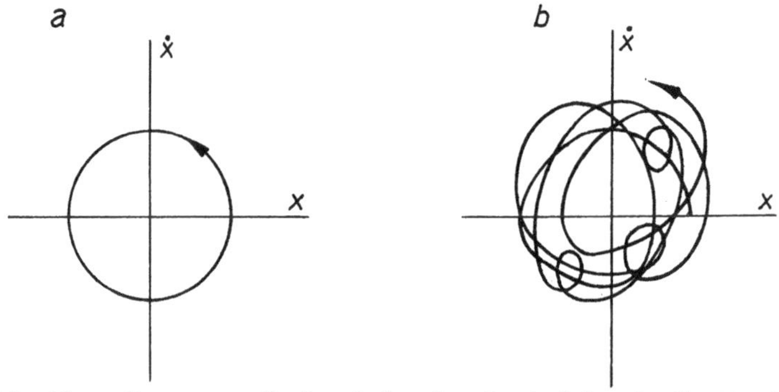

Fig. 7.5. Phase diagrams: a – limit cycle (regular attractor), b – chaotic attractor.

set up his "rules of three" in the following way (see Truesdell, 1968):

1. Observe the phenomenon and list the quantities having numerical magnitude that seems to influence it.
2. Set up such *linear* relations among pairs of these quantities as are not obviously contradicted by experience.
3. Propose these "rules of three" for trial by experiment.

From that the linearity seems to start his reign and Isaac Newton actually sealed it by his 2nd law in 1687:

"The change of motion is *proportional* to the motive force impressed, and it takes place along the right line in which that force is impressed."

Nevertheless, there were attempts to enhance the picture of Leonardo da Vinci. A good overview of linear and nonlinear stress relations is presented by Bell (1973). The stress-strain relations were most weakly nonlinear (Fig. 7.1b), but nevertheless in 1849 British Royal Commission recommended that henceforth Hooke's linear law of elasticity ($\sigma = A\varepsilon$) for iron in tension, compression, and flexure, should be replaced by $\sigma = A\varepsilon - B\varepsilon^2$ (Bell, 1973). To the knowledge of present author this recommendation has not been changed but engineering practice has very much relied on linear theory. Contemporary physics does not need the orders of highly distinguished commissions to accept nonlinearity because that is how the world is.

What does it mean then nowadays? When linear world is according to Maugin (1992b) uneventful then nonlinear world is full of interesting and unexpected events. In physics it means coupling of fields, solitons and other solitary waves, loss of stability, shock waves, fracture, chaotic heart beats, dissipative structures, fractals, etc., etc.

Two aspects should be stressed separately.

First, as said by Maugin (1992b)—it is not the nonlinearity itself that influences the outcome but the "possibility of compensation/balance or competition between nonlinearity and another characteristic property". This another characteristic property could be dispersion (leading to solitons), dissipation (shock waves, dissipative structures), forcing (chaotic regimes), etc.

Second, the nice predictability of ordered linear world may be lost in nonlinear world. The outcome is sensitive to the smallest cases in the initial data ("butterfly effect") and not predictable in a long run. What a blow to those who would expect weather prediction to reach 99.999 . . . % accuracy! In physical world it is not possible (Lorenz, 1963; Lighthill, 1986).

How nonlinearity enters in our models or understanding? Actually the simplest answer is that it is done by violating the Leonardo's 2nd rule: instead of linear relations we should "set up" nonlinear relations "among pairs of quantities". Even more, "pairs" may be too limited because more "quantities" may be interrelated than just "pairs" (the principle of equipresence). There are certainly restrictions also here, emphasized by the so-called Occam's razor: Do not multiply your working hypothesis more than is necessary.

In order to change linear models to nonlinear ones, even "nonlinearization technique" is proposed (Whitham, 1974) in context of improving the solutions for sonic boom.

And then we have to take the view of Scheid (1993). Paraphrasing him, we should say "Common sense usually so good at distinguishing between true and false fails *in nonlinear science.*"

Still, we may put a question: are there any bricks in the nonlinear world from which it could be built up? This question actually comes from the common sense, knowing for example that Fourier' modes are so important in the linear world. Again, the answer is twofold. First, there may be phenomena like solitons, wavelets, vortices, elementary hysteresis loops, etc., that are close to being fundamental units (Maugin, 1992b). But there is no possibility to put them together without interaction with what may bring new qualities into the structure of a system. So, from this viewpoint the answer is not known. Second, there is an important peculiarity—the complexity could arise from repeating simple nonlinear "units" (algorithms, structures, etc.)—the result is again completely different from the starting point. This is certainly the case of fractals (Mandelbrot, 1982). These problems are also analyzed in one of my earlier essays (Engelbrecht, 1993a).

So the question about bricks is possibly not correct. Said Georg Cantor: The art of asking the right questions . . . is more important than the art of solving them.

In order to establish the possible shortcomings on the rules of linear world, it is advisable to reconsider main assumptions again. I would not argue the Newton's 2nd law. It is clear that physical laws cannot be proved like the Pythagoras theorem but their validity has always been a question to philosophers of science. Cartwright (1984) has shown many cases when physical laws are correct only in *ceteris paribus.* In contemporary context, the Newton's 2nd law is the balance of momentum and more attention should be paid to its deepness rather than the imprint of proportionality. It must be stressed that the famous law of gravitation based on the 2nd law, is nonlinear (involving r^{-2}) and has given rise to the three-body problem—one of the best examples of nonlinear dynamics. H.Poincaré was first to realize that the calculation difficulties arising in the three-body problem meant something deeper, that was none other than unpredictability.

Returning to the balance of momentum, it reads in contemporary notation

$$\frac{\partial}{\partial t}p - \mathrm{div}T = 0\,, \tag{7.1}$$

where p is the physical momentum and T is the first Piola-Kirchhoff tensor. If tensor T includes nonlinear terms due to elastic moduli of higher orders or to nonlinearity of deformation tensor terms (Bland, 1969), then the final governing equation is nonlinear. This approach is not straightforward and needs several assumptions in addition. There is, however, a possibility to reformulate the balance of momentum in terms, reflecting the possible nonlinearity and other important effects more transparently. I follow here the ideas of Maugin (1992a), who worked with the Eshelby's tensor accounting for the stress field around inhomogeneities within a solid. Maugin understood that this tensor is actually the material tensorial flux associated with the conservation of the so-called pseudomomentum. That balance needs a source term accounting for material inhomogeneities to be included. As a result, the balance of the pseudomomentum reads

$$\frac{\partial}{\partial t}P - \operatorname{div} b = f^{\text{inh}}, \tag{7.2}$$

where P is the pseudomomentum, b is the dynamic version of the Eshelby's tensor, and f^{inh} is the material force accounting for elastic inhomogeneities. Leaving the details aside (for which the reader is referred to Maugin (1992a)), it is clear that such a description is more convenient to be used for inhomogeneous or structured media. As an example, let us take the Boussinesq equation for elastic crystals (Maugin, 1992a). If the Lagrangian density L is formulated as

$$L = \frac{1}{2}u_t - W(u_x, u_{xx}), \tag{7.3}$$

where u is the displacement with indices denoting the differentiation and

$$W = \frac{1}{2}\left(u_x^2 + \frac{2}{3}\varepsilon u_x^3 + \varepsilon\delta^2 u_{xx}^2\right), \quad \varepsilon = \text{const.}, \delta = \text{const.}, \tag{7.4}$$

then in the terms of the balance of pseudomomentum we have

$$P = -u_t u_x, \tag{7.5}$$

$$b = -\left(L + u_x\frac{\partial W}{\partial u_x} + 2u_{xx}\frac{\partial W}{\partial u_{xx}} - \left(u_x\frac{\partial W}{\partial u_{xx}}\right)_x\right), \tag{7.6}$$

$$f^{\text{inh}} = 0. \tag{7.7}$$

The Boussinesq equation itself is now easily derived

$$u_{tt} - u_{xx} - 2\varepsilon u_x u_{xx} - \varepsilon\delta^2 u_{xxxx} = 0 \tag{7.8}$$

including directly nonlinearity (the 3rd term) and dispersion (the 4th term). Such an approach has been applied to many problems like brittle fracture, ferromagnets, ferroelectrics, surface elastic solitons, etc. (see Maugin, 1992a and the references therein).

To sum up, there is nothing wrong with our basic laws, one should only get rid of the imprints they bear due to the long usage. Based on nonlinear effects, we are able to describe many physical effects in nature, like

solitons,
fractals,
dissipative structures,
pattern formation,
strange attractors,

hysteresis,
etc., etc.

One cannot describe sonic booms, nerve pulses, tsunamis, tidal bores, traffic jams, superconductors, fracture, combustion, a.o. without nonlinearity involved. Is it not fascinating that so many important phenomena are based on the negation? The importance of being nonlinear is nowadays widely accepted (West, 1985) because linear world is too simple.

Let me finish this essay by a somewhat heretical idea. If linear is simplification and nonlinearity reflects much better the reality of the world, then why determine the more fundamental notion by negation? Vice versa, a good word is needed to emphasize all these deep qualities of nonlinearities like the lack of superposition, unpredictability, coupling, quality of repetitive action, etc.

We know the balanced pairs of notions like high-dimensional vs low-dimensional or macroscopic and microscopic. We know that there are new words coined for phenomena not known before like dissipative structures or fractals. In the latter case Mandelbrot actually avoided negations. Actually we face the same sort of situation looking for a word to emphasize the basic values behind nonlinearity.

What we need first, is to overcome the usual "common sense" characterized by a simple one-way model of thought:

linear → nonlinear.

This negation of simplicity (linearity) seems nowadays to be carved in stone tablets and in minds of most people. Is it normal? Actually the one-way model of thought should stand in the different order:

nonlinear → linear (not nonlinear).

As the latter involves double negation then better is:

complex → noncomplex = simple.

Complex is a very slippery word and therefore not the best notion here. When writing these lines, a colleague of mine, H.Hiller said: "Nonlinearity is normality!". His chain looks like

normal (nonlinear) → simple (linear).

This is actually a balanced pair of notions to mark the deep qualities of normal (nonlinear) world. It should be stressed that normality is also a biological concept (Wachbroit, 1994) but then its negation—abnormality—means not functioning properly. The concept of normality used above is no negation, it can only be simplified. Some properties are certainly lost by simplification but whether simplification is abnormal or not, is another question. I would call simple simple just because a spade should be called a spade. So normal dynamics means as full description as possible including nonlinearity and other nonproperties, simple dynamics is just its simplified and linearized variant. It does not mean that all the problems in simple dynamics are solvable and solutions known but this is more related to semiotic complexity.

One should also not forget that our understanding of the world is always changing. Said Bachelard (1934): "In our research, we shall always meet the features of enlarging, inference, induction, generalization, synthesis and totality—this is all for the idea of novelty." In this sense the notion of normality is just progressing and does not need any "nons" to be added.

To sum up—it is quite normal that normal world is full of unexpected results. And certainly nonlinearity is neither a nonproperty nor nonsense but normality.

7.2. BEAUTIFUL DYNAMICS

> *It is impossible to follow the march of one of the great theories of physics without being charmed by the beauty of such a construction.*
>
> *Pierre Duhem, 1906*

Scientific methods imply that all the notions used for the explanation of certain ideas should be clearly defined. Dynamics is just one chapter in this unfinished book called Science, hence I should start from the definition of science. The Chambers (1982) says that science is "the ordered arrangement of ascertained knowledge, including the methods by which such knowledge is extended and the criteria by which its truth is tested". If something is ordered, then beside the information, probably embedded into the ordering, there should also be some aesthetical value in it. The question is whether this means beauty or not. Again dictionaries should be consulted. The Collins (1989) explains beauty as "the combination of all the qualities of a person or thing that delight the senses and please the mind". As a consequence, the adjective *beautiful* is determined like possessing beauty or aesthetically pleasing. However, the Shorter Oxford English Dictionary (1962) says that beautiful is pleasing to the sense or intellect. This seems to be an excellent justification for the title of this essay.

Without any doubt, the contemporary nonlinear dynamics is pleasing to the intellect. The world around us is much richer indeed when we are familiar with solitons, strange attractors, fractals and other nonlinear phenomena. This essay is about the dynamics of solitons that certainly has an interesting history, extraordinary scientific depth and remarkable beauty in the sense explained above. Having said that, I have to prove it or at least provide examples to justify what has been said. That is the topic of this essay.

The reason of collecting some general thoughts on solitons just now is simple: in 1995 the scientific community celebrated the centennial of the publication of the famous paper by Korteweg and de Vries (1895) on a new type of stationary waves. That has been one of the most important milestones in soliton dynamics or solitonics, the value of which is hard to overestimate.

There were some other important milestones before and after 1895 and only in 1965 it became clear that this narrow path of science had to be converted into a highway. Contrary to what happens in everyday traffic, this highway has opened the beauty and richness of nature in a wide scale. We pass now these milestones step by step.

7.2.1. Milestone 1

Many scientific discoveries begin "with the experimental or observational isolation of an anomaly", says Kuhn (1977). The discovery of solitons began in 1834 on the Union Canal near Edinburgh. This place is not far from the contemporary campus of Heriot-Watt University built more than 100 years later. It is not surprising that now a strong group of soliton specialists is working there. The observation of an anomaly was reported by John Scott Russel, a naval engineer. His description is now one of the most cited paragraphs in solitonics and, as many others, I cannot resist repeating his words: "I believe I shall best introduce this phenomenon by describing the circumstances of my own first acquaintance with it. I was observing the motion of a boat which was rapidly drawn along a narrow channel by a pair of horses, when the boat suddenly stopped—not so the mass of water in the channel which it had put in motion; it accumulated round the prow of the vessel in a state of violent agitation, then suddenly leaving it behind, rolled forward with great velocity, assuming the form of a large solitary elevation, a rounded, smooth and well-defined heap of water, which continued its course along the channel apparently without change of form or diminuation of speed. I followed it on horseback, and overtook it still rolling on at a rate of some eight or nine miles an hour, preserving its original figure some thirty feet long and a foot to a foot and a half in height. Its height gradually diminished, and after a chase of one or two miles I lost it in the winding of the channel. Such, in the month of August 1843, was my first chance interview with that singular and beautiful phenomenon which I have called the Wave of Translation, a name which it now very generally bears." (1844)

Few scientific discoveries have been described with such verbal exactness and beauty. Scott Russel was convinced that what he had observed was an important phenomenon. He reported later (1865): "No one before had fancied a solitary wave as a possible thing."

The next stage of a scientific discovery is the period when the discoverer and his supporters try to prove the existence of the phenomenon and quite often some other group tries to disprove it (Kuhn, 1977).

To be more exact, Scott Russel's results could be formulated as follows (Ablowitz and Clarkson, 1991):

(i) he observed long, shallow water waves of a special permanent form and concluded that they exist;

(ii) he determined the speed of the propagation of these waves depending on the amplitude of the wave (and on the depth of the channel).

For several decades there were debates in the scientific community until Boussinesq (1871a) and Rayleigh (1876) derived approximate descriptions of the solitary wave as the hyperbolic secant square for the free water surface. Lord Rayleigh actually gave an excellent account on the quest for solitary waves from Scott Russel to Airy, Stokes, and Earnshaw, deriving himself a solitary wave solution. Boussinesq (1871a, 1871b), in turn, paid more attention to equations that are nowadays also widely used as the two-wave equations. The debates before these events where not very quiet because the scientific community (Stokes, among the others) did not believe in the possible existence of a "hump" wave without balancing it with another polarity. Probably most researchers simply kept Leonardo da Vinci's opinion (Truesdell, 1968): "By so much higher are the waves of the sea than the ordinary height of the surface of its water by that much lower are the bottoms of the valleys interposed between these waves". Alas, it is not always true!

Now, looking back and knowing much more than scientists of the 19th century, it is easy to realize that the discussion was just about the additional, properly scaled terms to the wave equation. This discussion is explicitly explained by Sander and Hutter (1991), using an explanatory scheme in $\varepsilon = h^2/\lambda^2$ and $\sigma = \eta/h$ plane. Here h is the depth of water, λ – the wavelength and η – the amplitude of the wave.

Thanks to these studies, the path was much wider in the second half of the 19th century. Nevertheless, counting from Scott Russel's discovery, only after 60 years the scientific community reached

7.2.2. *Milestone 2*

This is a special one to mark in 1995. The story is not so picturesque as the previous one, but just a story about a scientific paper. Diederik Johannes Korteweg and Gustav de Vries, a professor and a doctoral student at the University of Amsterdam, published in 1895 their paper entitled "On the change of form of long waves advancing in a rectangular canal, and on a new type of long stationary waves". It is worth to notice that (i) the title contains the words "change of form" referring to dynamics, i.e. to an equation governing the change and (ii) "a new type" means cnoidal waves. The equation itself is able to produce solitary as well as cnoidal waves and in its original form it is written

$$\frac{\partial \eta}{\partial t} = \frac{3}{2}\sqrt{\frac{g}{l}} \cdot \frac{\partial \left(\frac{1}{2}\eta^2 + \frac{2}{3}\alpha\eta + \frac{1}{3}\sigma \dfrac{\partial^2 \eta}{\partial x^2} \right)}{\partial x} \tag{7.9}$$

where η is the elevation of the surface, l is the depth of the canal, g is the gravity acceleration, and α, σ are the constants, the latter depending on capillary tension. Equation (7.9) possesses a stationary solution

$$\eta = h \operatorname{sech}^2 x \sqrt{\frac{h}{4\sigma}} \tag{7.10}$$

to which Korteweg and de Vries referred as "well-known" after Boussinesq (1871a) and Rayleigh (1876).

After the scale and variable transformations Eq. (7.9) or the Korteweg-de Vries (KdV) equation can be represented as

$$u_t + uu_x + \mu u_{xxx} = 0\,, \quad \mu = \text{const.}\,, \tag{7.11}$$

where indices denote the differentiation and μ stands for the dispersion parameter. The KdV equation and its modifications form nowadays a backbone of nonlinear dynamics. However, probably Korteweg and de Vries did not shout "Eureka", because their interest was just modelling waves in shallow water. Stressing Eq. (7.9) as a "very important equation" they probably were more glad to describe cnoidal waves. The paper is written on the basis of the doctoral dissertation of de Vries (the promotion took place in 1894) and not much is known about the author's scientific activities later on, except for his two other papers on cyclones. Probably he was quite happy with his schoolteacher's life. Korteweg, however, is also known because of his other studies. His doctoral dissertation under the supervision of J.D. van der Waals, was on the velocity of waves in elastic tubes filled with a fluid (Dict. Sci. Bibl., 1973). In the theory of blood flow in arteries, this velocity is now known as Moens-Korteweg velocity of pressure waves, related to the radial motion of the artery's wall. It is interesting to note that van der Waals paid a tribute to his former student in his Nobel Prize Lecture.

Why is the KdV equation so special from the contemporary viewpoint? First, it describes the propagation of solitary waves; second, it shows the relation of solitary waves to harmonic waves; third, the KdV equation is the simplest nonclassical partial differential equation (p.d.e.) possessing (Ablowitz and Clarkson, 1991)

- the minimum number of independent variables (2);
- the lowest order of the derivative not considered classically (3);
- the fewest terms of that order (1);
- the simplest such term (an unmixed derivative);
- the smallest number of terms (1) containing the other derivative, which is of the first order;
- the simplest structure for this term (linear);
- the simplest additional term to make the equation nonlinear (quadratic).

Notice that the case with u_{xx} as the highest derivative in a similar p.d.e. is directly related to the classical heat conduction.

Science progressed in time. Korteweg and de Vries served a footnote about 10 years later in the third edition of the fundamental treatise on hydrodynamics by Lamb (1906), who said, "... as the title indicates, the paper includes an examination of the manner in which the wave profile is changing in an instant, if the conditions of permanency of type are not satisfied." It took 70 whole years to reach

7.2.3. Milestone 3

Actually this milestone has two facets, one marking the path coming from Scott Russel and Korteweg & de Vries, the other—from Fermi, Pasta, and Ulam (1955). The Fermi-Pasta-Ulam (FPU) problem was actually related to the process of relaxation to thermal equilibrium in lattices. The lattice they studied, was one-dimensional, nonlinear, and the equations of motion involved nearest interactions only

$$d^2x_k/dt^2 = F(x_{k+1} - x_k) - F(x_k - x_{k-1}), \quad k = 1, 2, \ldots, N, \qquad (7.12)$$

where x_k is the displacement of the kth particle and F is the nonlinear force (quadratic, cubic, ...) while $N = 32$. Their idea was to study how this system relaxes to equilibrium when the energy is initially put into the lowest mode. What they found was that the energy was not distributed over all 32 modes of this system but was shared by only several modes. Even more, after a certain time, energy returned to the first mode! It is worth to notice that they used the MANIAC I computer at Los Alamos. So the history of solitary waves is closely related to the history of computers.

The FPU problem initiated the studies of Zabusky and Kruskal (1965). They took $x_k(t) = kh + y_k(t)$ where h is the free-lattice constant and y_k the displacement from the equilibrium. Their simple but ingenious idea was to replace the lattice by a continuum. For this purpose, $y_k(t)$ was replaced by $y_k(x,t)$ using the substitution

$$y_{k+1}(t) = y(x,t) \pm hy_x + \frac{1}{2}h^2 y_{xx} + \dots, \tag{7.13}$$

where index x denotes the differentiation. Retaining the terms through $O(h^4)$, the outcome was the KdV equation!

The numerical integration of the KdV equation (possible at that time!) produced a remarkable result. From a periodic initial condition (cosine function) a series of pulses emerged, each of which was a solitary wave. Figure 7.6 taken from Zabusky and Kruskal (1965) is probably the most reproduced (after the quotation by Scott Russel) result in the history of solitary waves. Eight pulses, as seen in Fig. 7.6, have emerged for a certain parameter μ (see Eq. (7.11)). Alas, as the solitary wave speed depends upon the amplitude (guessed already by Scott Russel), then a larger pulse always propagates faster than a smaller. It means that a larger pulse always overtakes a preceding smaller one. The first surprise was that the overtaking, i.e. the interaction did not change the amplitude of both pulses and resulted only in a certain phaseshift. The second surprise was that after a certain time the initial state (periodic initial condition) was restored. The pulses (solitary waves) behaved like particles! Because of such behaviour Zabusky and Kruskal (1965) coined the name *soliton* (c.f. electron, photon, etc.). In the title of their paper, they marked "solitons" in inverted commas and the definition is the simplest that one can only think of. They namely stated that "... each such "solitary-wave pulse" or "soliton" begins to move ..." The aim of their research was certainly not to discover solitons but to cast more light on the FPU problem. Zabusky and Kruskal (1965) concluded in their final sentence that "because the solitons are remarkably stable entities, preserving their identity through numerous interactions, one would expect thermalization (...) only after extremely long times, if ever".

It is worth to mention that Boussinesq (1871b) has used the word *intumescence* for a moving disturbance of the free surface. This term, however, is not often used nowadays.

That was the starting point of the soliton era or the quest for localized solutions in spatially extended systems. Said Roger Cotes in 1713: "The gates are now set open." These words originate from the preface to *Principia* but could also be used here to emphasize the success of soliton theory.

7.2.4. Contemporary highway

Some definitions are needed now. In 1995, we know much more than the scientists who have marked the milestones listed above. Not only the KdV equation, but many other important equations used in mathematical physics, possess soliton-type solutions, with varying properties. Hence, the solitons should be related to the equations resp. physical processes (Engelbrecht, 1990) and the soliton discussed above is actually the KdV soliton. It results from the balance between quadratic nonlinearity $(1/2(u^2)_x = uu_x)$ and cubic dispersion (μu_{xxx}).

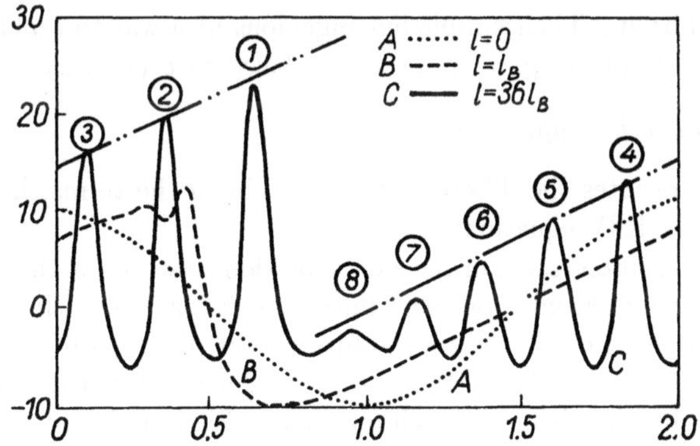

Fig. 7.6. Solution of the KdV equation at different times (after Zabusky and Kruskal, 1965).

A general definition of a soliton is the following. Soliton

(i) is a wave of permanent form;
(ii) is localized in space;
(iii) preserves its structure and velocity after an interaction with another soliton.

A KdV soliton has an additional property: its speed depends upon the amplitude. It is interesting that in quantum mechanics the statement is usually inverse: the amplitude depends upon the speed. Other solitons may not have this property.

Widely used is nowadays the normalized KdV equation

$$U_t - 6UU_x + U_{xxx} = 0 \tag{7.14}$$

with its solution

$$U = -2k_i^2 \mathrm{sech}^2 \left[k_i(x - 4k_i^2 t + x_i)\right] , \tag{7.15}$$

where k_i is an eigenvalue of the corresponding Schrödinger equation with the initial condition to Eq. (7.14) as a potential, and x_i is a phase shift.

The aim of this essay is neither to give the overwhelming history of soliton theory nor to present a review of the results. However, some highlights of theory should be mentioned that have directly been generated by those basic ideas discussed above:

- direct generalization of the Fourier transform in order to solve nonlinear equations; the method is called inverse scattering transform (IST) and has many modifications (particularly, solution (7.15) is derived using IST);
- theory of Toda lattice showing that solitons are the signature of integrable systems;
- Hirota, Painlevé, Bäcklund, and other methods,

etc., etc.

The soliton structures appear in many areas. Manton (1994) has shown

(i) the one-dimensional (1D) solitons: waves in shallow water, signals in optical fibres;

(ii) the two-dimensional (2D) solitons: magnetic flux domains in superconductors, vortex-antivortex pairs in fluids;

(iii) the three-dimensional (3D) solitons: magnetic monopoles in gauge theories, skyrmions, i.e. soliton models for protons and neutrons.

Solitons may be the mechanism for energy transmission along the alpha-helix protein molecules (Davydov solitons), they may arise in gas-fluidized beds (chemical industry, food processing), they may be of importance in computer hardware (Josephson junctions), etc. Much can be found in the excellent reviews (Sander and Hutter, 1991; Jeffrey and Kakutani, 1972; Scott et al., 1973; Miura, 1976) and books (Ablowitz and Clarkson, 1991; Lonngren and Scott, 1978; Newell, 1985; Davydov, 1985; Toda, 1989; Fordy, 1990; Fokas and Zakharov, 1993) which are a few among the others.

Now the basic question again:

7.2.5. Where is beauty?

The variety of applications and the deepness of the theory are hallmarks of solitonics. We might also ask whether this success has something to do with the beauty of the phenomenon, or with the simplicity of basic concepts, or with something else.

Kuhn (1977) lists the characteristics of a good scientific theory: accuracy, consistence, scope, simplicity, and fruitfulness. I am tempted to give some of his ideas in their full form:

- "... a theory should be accurate: ... consequences deducible from a theory should be in demonstrated agreement with the results of existing experiments and observations";
- "... a theory should be consistent ... with other currently accepted theories applicable to related aspects of nature";
- "... a theory's consequences should extend far beyond the particular observations, laws, or subtheories it was initially designed to explain";
- a theory "... should be simple, bringing order to phenomena that in its absence would be individually isolated ...";
- "... a theory should be fruitful of new research findings ..., disclose new phenomena or previously unnoted relationships among those already known".

What a precise description of soliton theory! Indeed, it has stimulated a tremendous amount of theoretical work in mathematics and related fields, and it has an enormously wide range of applications (see above). The simplicity may be the point of arguments. The IST may seem complicated but it is based on simple ideas and has more semiotic complexity rather than ontological complexity (Kampus and Czanyi, 1987). Said Fokas and Zakharov (1993): "Let KdV describe the propagation of a water wave and suppose this wave is frozen at a given instant of time. By bombarding this water wave with quantum particles, one can reconstruct its shape from knowledge of how these particles scatter." This is the simple idea behind the IST.

Is simple beautiful? Scott Russel was the first to recognize the beauty of a solitary wave. He said 1865: "This is a most beautiful and extraordinary phenomenon: the first day I saw it was the happiest day of my life." Having seen solitons in a laboratory at the University of Grenoble, I take Russel's words very seriously. These solitons in Grenoble were created in a long ($\sim$ 30 m) canal by an impact-type excitation at the end of the canal. The solitons indeed propagated, overtook each other, scattered—just in a simple way predicted by the theory. Beautiful they were, indeed!

Returning to the definition given at the beginning of this essay, intellect was also mentioned in the context of beauty (Shorter Oxford Dict., 1962). From the intellectual viewpoint, the IST is a beautiful method, solving the nonlinear equations by a sequence of linear operations resp. equations (frozen wave and bombarding (Fokas and Sakharov, 1993)).

Great scientists whose destination was not only to explain a certain physical phenomenon but to go into the depth of Nature secrets, have stressed the beauty of scientific theories. Said Paul Dirac: "A physical law must possess mathematical beauty," that for him was an intrinsic feature of nature (Hovis and Kragh, 1993). Much earlier Pierre Duhem (1914, cited after 1954), another priest of science, said: "Order wherever it reigns, brings beauty with it. Theory not only renders the group of physical laws it represents easier to handle, more convenient, and more useful, but also more beautiful."

Says the first Encyclopedia Britannica (1769): "... order and simplicity contributes greatly to enhance the beauty of objects that surrounds us ..." Again, in soliton theory, the simplicity of governing principles involved permits to deduce a wide range of conclusions.

The same source mentions also the beauty of circles and squares. Quite surprisingly, modern Encyclopedia Britannica does not pay any attention to beauty. The entry of beauty culture of the 1964 edition refers to cosmetics and cosmetology... Even more, the New Encyclopedia Britannica of 1991 has entries only on plants, the names of which contain the word "beauty". This seems not to be in line with the motto of this respectable source: "Let knowledge grow from more to more and thus be human life enriched". Not denying the human beauty, I think that the beauty of nature is far more rich than the beauty of just one of its species.

The stunning beauty of fractals, emphasizing order and self-similarity of nature has been widely accepted because of computer graphics and visualization (Peitgen and Richter, 1986). For soliton specialists, the timeplots showing the evolution or emerging of solitons are without any doubt also visually beautiful. In Fig. 7.7, for example, a process of soliton formation and interaction from a harmonic input is shown. Nine solitons emerge and interact with each other in harmony creating a beautiful ordered pattern. Nonspecialists should be asked whether this figure is good to look at. An artist might ask about the scientist's role in creating such a vision. The answer goes back to Korteweg and de Vries who have implicitly foreseen the beauty of the solutions to the equation that now bears their name.

7.2.6. Endnote

Without any doubt, solitonics is even so a Pandora's box. Just two recent examples might cast some light on possible new discoveries from the box.

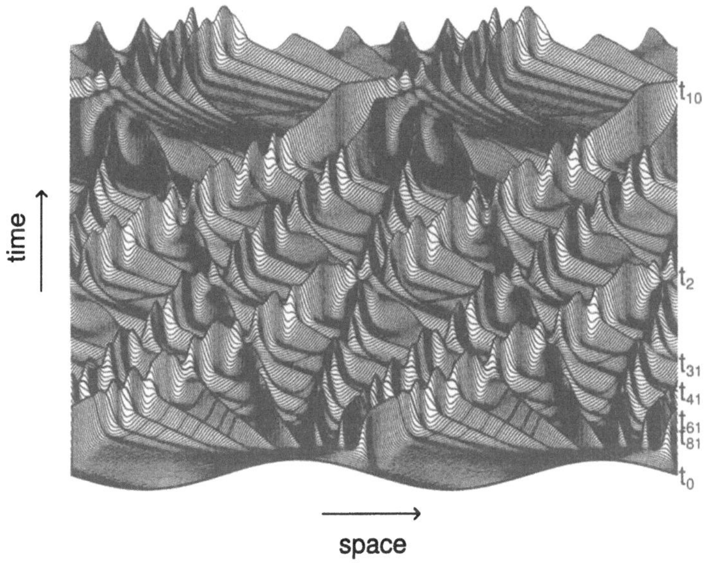

Fig. 7.7. Time slices for soliton formation from a harmonic initial excitation (calculated by A.Salupere).

The first example considers solitary waves with compact support. The KdV solitons can be derived from cnoidal waves letting period $T \to \infty$, i.e. a KdV soliton has an infinite span despite being highly localized in space. Rosenau and Hyman (1993) have introduced a family, as they call, of fully nonlinear KdV equations $K(m,n)$

$$u_t + (u^m)_x + (u^n)_{xxx} = 0\,, \quad m > 0\,, \quad 1 < n \leq 3\,. \tag{7.16}$$

They have shown for certain $K(m,n)$, particularly for $K(2,2)$ that the solutions to these equations have compact support, i.e. vanish outside a finite core region. These solutions are called compactons. Similarly to KdV solitons, compactons reemerge after colliding with their own kind but the collision is accompanied by the birth of a compact ripple. It seems another promising path to follow while the compact support is physically understandable and attractive. In Rosenau and Hyman (1993), the basic equation was reported to describe the formation of liquid drops. What could be other nonlinear structures governed by equations $K(m,n)$? Can, for example, compactons be used to pack tightly signals in optical fibres or can they be used in the field theory? The latter question put by Rosenau and Hyman (1993) is really intriguing.

The second example is from a more technical side. An idea of a "soliton tube" is proposed by Sugimoto (1993) to build a tunnel for high-speed (540 km/h) trains in Japan. This tunnel should have a main passage and many cavities arranged along both the axial and circumferential directions of the tunnel. The role of these cavities is to create dispersion together with additional damping and then instead of a shock wave (sonic boom) ahead a train acoustic solitons propagate governed by the KdV equation. The shock-free tunnels solve essential environmental noise problems. This is without any doubt an excellent idea based on knowing the role of dispersion and creating then a technical soliton-bearing system.

Within many scientific areas, soliton dynamics has been "... more eminently excellent than the rest of that with which it is united", according to Johnson and his famous dictionary (1755). These words were under the entry *beauty* and as an example, Bacon's words have been added: "The best part of beauty is that which a picture cannot express." I cannot agree with these words more, and as these lines have been written in Cambridge, I am tempted to finish with paraphrasing the words of Sir Isaac Newton, a Fellow of Trinity College: only a few pretty pebbles and shells on the seashore of the great and beautiful ocean of nature have been described.

7.3. COMPLEXITY AND SIMPLICITY

An essay is by the definition a composition dealing with a subject from a personal point of view and without attempting completeness.

B.Mandelbrot

When I told a colleague of mine that I would like to write an essay on complexity and simplicity, to Nikolai Alumäe's memorial issue, then he, a well-read scientist with a broad outlook on science and life, responded: "It is not an easy task." I cannot agree with him more, but then this is exactly what Nikolai Alumäe would have liked. The problem on one's desk should always be intriguing, he used to say, and I dedicate this essay to the memory of my Teacher.

7.3.1. Of initial ideas

The eternal rivalry between complexity and simplicity has always been an intriguing problem to mankind. As our knowledge has been increasing, many seemingly complex phenomena have turned out to be rather simple when the reasons behind them have been cleared up. Take, for example, the motion of planets that ages ago needed special spheres and other geometrical structures to explain their motion with respect to the Earth. Yet, as soon as the heliocentric system was accepted by the general public, everything became clear and simple. So, the first lesson is that, given the reasons, seemingly complex things become rather simple.

However, the world around us is not only a hide-and-seek game—there are simple phenomena and there are complex phenomena in it. The knowledge that God does not play dice, as Albert Einstein has said, is pleasing but not very helpful. As usual, we first need clear definitions, at least for the working purpose. According to Chambers (1961), *simple* is that which consists of one thing or element, *complex* is what is composed of more than one or of many parts. Taking these notions as basic, we later enlarge them intuitively, because, as a matter of fact, the categories involved are rather abstract.

Here, we shall start from examples demonstrating simplicity and/or complexity as we understand these phenomena (properties); and it is of fundamental importance to distinguish between the two, because contemporary theories tend to be very complex, so simplification seems to be a natural way to solve problems. Here lies a certain risk. A.Toffler mentioned in the Preface to I.Prigogine's and I.Stengers' monograph (1984): "Dissection is one of the most highly developed skills in contemporary western civilization: the split-up of problems into their smallest possible components. We are good at it. So good, we often forget to put the pieces back together again".

From this, an intriguing problem arises—how to put the pieces back again, i.e. how to build up complexity from simplicity. Below, we would like to analyze some steps on this long and difficult way, which may serve as a backbone for further research. And there are a lot of questions to be answered before complexity is understood more profoundly.

7.3.2. Of simplicity

According to what is said above, simple consists of one thing or element, which is a rather open definition. Here we have to rely more on the intuition that permits us to enlarge these notions without violating the basic idea. Simplicity in mathematics is related to simple rules, in physics—to linear causality like "if this is given then that follows", etc. Let us proceed with the aid of exemplification.

We are faced with simple rules quite often in our everyday practice. Engineers widely use Hooke's law stating that stress and deformation are proportional, Ohm's law—current and voltage are proportional, etc. These can be considered as simple laws emphasizing simple instantaneous effects—the larger one variable, the larger also the dependent variable. The cause and the result are nicely related to each other and proved to be correct in ever so many everyday examples. There are cases where only a simple model has opened the eyes of mankind to understand the rules of Nature. A brilliant example is the planetary model of the atom based on the ideas of E.Rutherford and N.Bohr.

There are two questions now: (i) is simple sometimes just simplified and (ii) what is the difference between simple and simple. The latter question seems to be a tautology but its essence is to widen the notion of simplicity.

It is quite clear that simple linear laws of physics are applicable only under special conditions (*ceteris paribus*). So, stress is proportional to deformation only if deformation is small. The real world is nonlinear and Hooke's law is only a simplified version of reality. This is easily understood when analysing the properties of potential energy. In Fig. 7.8 potential energy U is depicted as a function of distance r between the atoms. Around the equilibrium point O, the approximation by a linear function is justified, otherwise the energy function is essentially nonlinear (higher than quadratic).

In order to answer the second question, we need a certain measure. Not going into details, an intuitive approach may be used here. In a physicist's jargon, the word *deep* is used to describe the problems "that would not give way without long looks into Universe's bowels..." (Gleick, 1988). Simplicity may also be deep. For

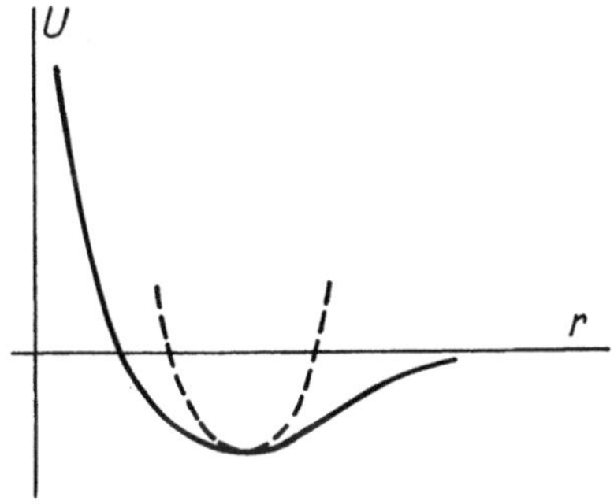

Fig. 7.8. Potential energy as a function of distance between atoms.

example, we can turn to thermodynamics. Such notions as heat, energy, entropy and dissipation are simple but deep. Let us take the first of them—heat. This notion was introduced by J.J.Fourier in 1811. The essence of his idea is surprisingly simple and elegant—heat flow is proportional to the gradient of temperature. Said Prigogine and Stengers (1984): "The simplicity of Fourier's mathematical description of heat propagation stands in sharp contrast to the complexity of matter considered from the molecular point of view". However, the deepness of this new notion has enabled to build up thermodynamics based on simple principles. According to R.Feynmann, this is quite usual in Nature, which is actually a huge chess game: each move follows simple rules (Lee, 1992). Nowadays we may ask, is such an approach still efficient. Some considerations concerning answers to this question will be given later.

7.3.3. Of complexity

The initial definition of complex (see above) is also certainly rather naive. One could ask, for example, what is the minimum of complexity involved that allows us to use the notion of complex. Intuitively we would like to add some other properties to the initial notion in order to grasp its fundamental essence.

According to Lee (1992), contemporary understanding of complexity is closely related to Gödel's paper (1931), "focussing on undecidability and the allied concepts of uncertainty and complexity". Nowadays our vision of Nature is undergoing radical changes towards the multiple, the temporal, and the complex.

Where does complexity come into the game? One should first distinguish between the structural and functional complexities (Nicolis, 1986). The first depends on the number of interacting subunits and the second on the length of the algorithm needed to describe the entire behaviour of the system. Complexity could also arise from the interaction between the system and its observer (Casti and Karlquist, 1986). Nicolis and Prigogine (1989) stress the importance of natural complexity seen as part of everyday experience closely related to nonequilibrium states, nonlinear dynamical systems, predictability, and self-organization. Packel and Traub (1987) show the importance of computational complexity. This short list may serve as evidence of many facets of complexity. Here, we concentrate our attention on natural complexity.

Fig. 7.9. The Mandelbrot set.

Again, as above, let us discuss some examples. When B.Mandelbrot in the seventies started to investigate a "simple" quadratic map

$$z_{n+1} = z_n^2 + c \tag{7.17}$$

with z_n, c complex, the question to him was, "Do you really expect to find anything new?" As we know now, the Mandelbrot set M (the domain of convergence of map (7.17) in the complex plane) is a very complicated structure, indeed (Fig. 7.9). It has rightly been described as the most complex mathematical shape ever invented (Stewart, 1989). Even more, the Mandelbrot set is another important milestone in the theory of complexity because from it the notion of fractals has started (Mandelbrot, 1977; Peitgen and Richter, 1986). According to Nicolis and Prigogine (1989), the complexity of the Mandelbrot set is certainly natural. We shall later return to this example when we discuss the rules of creating complexity.

Generally speaking, fractals are complex structures like coastlines, rivers, blood vessel systems, neural networks, cauliflowers, diesel soots, polymer grains, colloid aggregates—this list could be prolonged, of course (see Kaye, 1989; Bunde and Havlin, 1991). Let us cite Mandelbrot (1982): "Clouds are not spheres, mountains are not cones, coastlines are not circles, and bark is not smooth, nor does lightning travel in a straight line... Nature exhibits not simply a higher degree but an altogether different level of complexity". Fractal geometry is progressing (Peitgen and Saupe, 1988), discovering more and more rules in Nature.

Fractals depict the complexity of irregular and fragmented shapes in Nature. Besides them, there are many more complex structures and phenomena. In dynamics, there is a notion of complex dynamical systems built of units that are themselves simpler systems (Thompson and Stewart, 1986). The behaviour of dynamical systems in multidimensional phase spaces may be complex (but not fractal!) (Thompson and Stewart, 1986; Moon, 1987). The ecological and human organizations are usually complex (Prigogine and Stenger, 1984). Or just an example in physics—gas as a purely deterministic aggregate of moving molecules that obey precise dynamical laws (Stewart, 1989), and again it is a complex structure. The plane- and spacefilling tiling patterns are used not only for covering a floor or a wall with ceramic tiles but are of the utmost importance in crystallography (Peterson, 1988). The Penrose decagon tiling is shown in Fig. 7.10. What fascinates us in tiling (tessellation) is that Maurits Escher, the famous Dutch artist, has discovered many exact laws of tiling by just studying symmetry from an artist's viewpoint (Schattschneider, 1990). An example of his tiling is shown in Fig. 7.11.

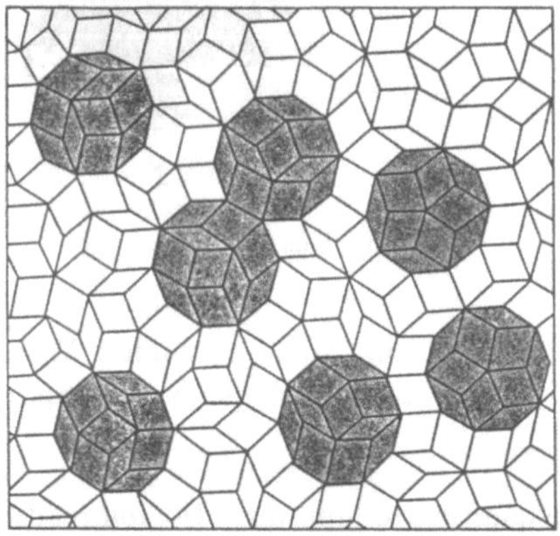

Fig. 7.10. Penrose decagon tiling (after Peterson, 1988).

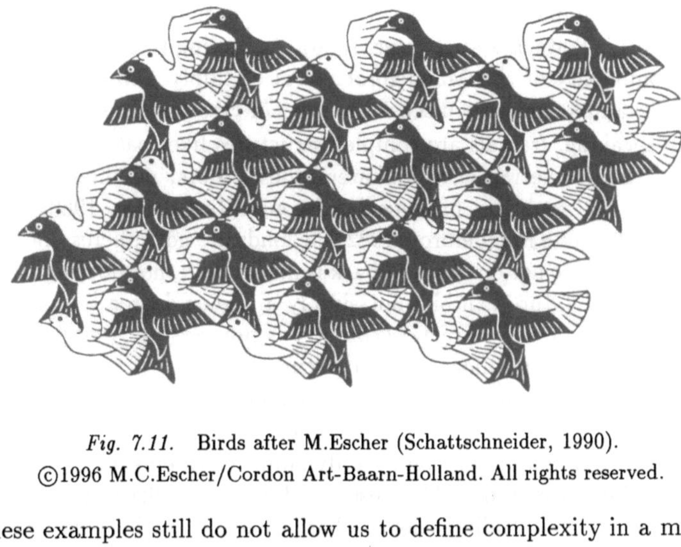

Fig. 7.11. Birds after M.Escher (Schattschneider, 1990).

All these examples still do not allow us to define complexity in a more definite way. We may list some of the characteristics of complexity:

– we do not easily understand the topological, geometrical a.o. properties of a complex structure;

– we do not ad hoc understand how a complex system is likely to respond to a given excitation (change of conditions).

7.3.4. Of creating complexity

Naive argumentation puts some simple things together in order to get complex structures. Actually, this basic idea is realized in a more sophisticated way by means of the following procedures:

– building hierarchies;
– using recursive algorithms;
– using coupling between structural elements;
– taking into account memory effects.

This list is by no means complete.

There is an important point to take notice of whatever the procedure to create complexity—the outcome may depend on a set of parameters. Let us take for example recursive algorithm in the form of a quadratic map

$$x_{n+1} = c - x_n^2 \qquad (7.18)$$

which, contrary to (7.17), is used in the real domain only. If $c < 0.7$, then it leads to a stable stationary point after a series of iterations (recurrent usage of map (7.18)). This is shown in Fig. 7.12a. Intuitively we may call this result simple. If $c \sim 0.78$, then a bifurcation appears, and instead of one stable point two stable points appear (Fig. 7.12b). Let us leave it open whether this result is still simple or not and proceed with the calculations for other values of c. In Fig. 7.12c, the results for $1.72 < c < 1.82$ are shown. These results are complex without any doubt! There are regions where no stable points occur, there is a window where a certain regular structure appears, etc. Actually, we have demonstrated here the celebrated period doubling scenario governed by Feigenbaum numbers (Feigenbaum, 1980). These universal numbers govern the transition to chaos and do not depend on the algorithms but only upon the fact that period doubling (flip bifurcation (Thompson and Stewart, 1986) appears (Stewart, 1989; Moon, 1987; Feigenbaum, 1980). This gives evidence of the existence of a certain universality in complex structures.

Returning to the Mandelbrot set (Fig. 7.9), which is related to the same quadratic map (7.18) but in the complex plane, we discover a striking quality of complexity—its richness. We only need to blow-up the structure, i.e. to move to ever smaller scale. This is a real Mathematical Zoo that can then be discovered with intricate beauty and endless variety: seahorses, scrolls, whirlpools, lumps, sprouts, burgeoning cacti, thin snakes, coils, insect-like blobs, zigzag lightnings, etc. (Stewart, 1989). The reader is referred to coloured plates, say in (Peitgen and Saupe, 1988), in order to get some impression about the beauty of complexity. Lewis Carroll would have put them into his Jabberwocky, had he, an Oxford don and mathematician, only known about their existence.

Another example of recursive algorithms could be the construction of "monster" curves and structures (Peitgen et al., 1992). The well-known Koch curve is depicted in Fig. 7.13 with its recursive shapes. We start from a straight line called initiator

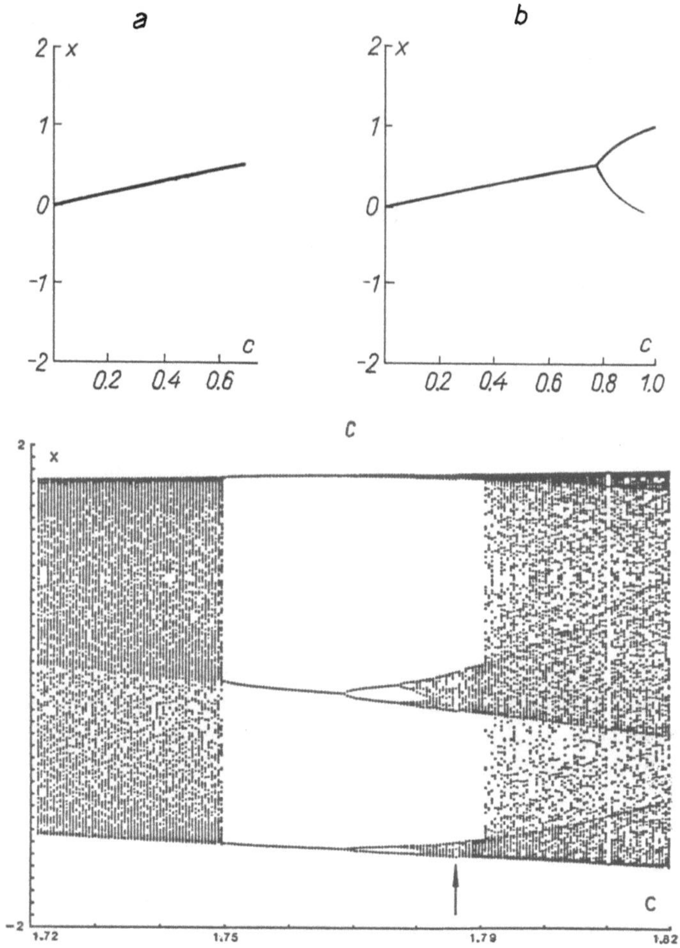

Fig. 7.12. Behaviour of stable points of quadratic map $x_{n+1} = c - x_n^2$: a) $0 < c < 0.7$; b) $0 < c < 1.0$; c) $1.72 < c < 1.82$.

(step 0). We partition it into three equal parts and replace then the middle third by an equilateral triangle, and take away its basis. This is the generator (step 1) which is then used repeatedly (in Fig. 7.13 up to 4 steps). The idea how to construct a Pythagorean tree is shown in Fig. 7.14, where an initiator is a square to which a right isosceles triangle is attached to one of its sides, after which two squares along the free sides of the triangle are attached, etc. A product where triangles are non isosceles, is shown in Fig. 7.15. A cut-off of a cauliflower, isn't it?

The notion of hierarchical systems is used in the information theory (Nicolis, 1986) and neural network architecture (Clark, 1989). For example, McCulloch-Pitts neuron is based on a finite number of symbolic expressions reflecting the behaviour of a single neuron. The combination of simple arithmetics, classification and storage of information, and recursive operation permits to build up hierarchical networks as

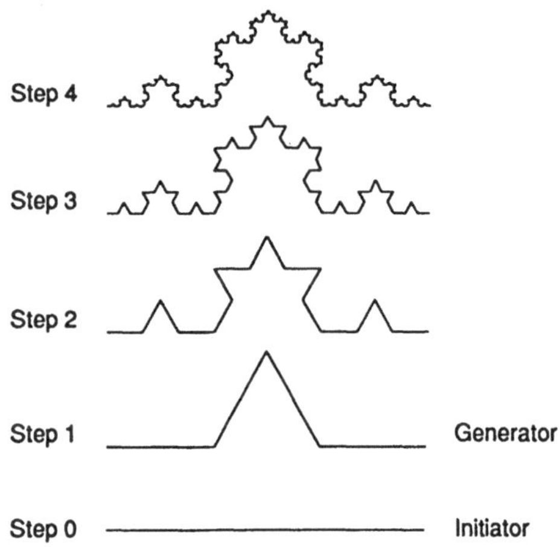

Fig. 7.13. Construction of the Koch curve.

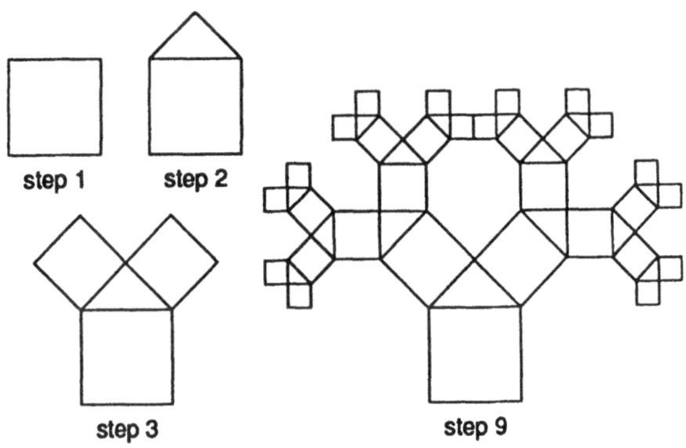

Fig. 7.14. A Pythagorean tree.

finite-state sequential machines. Here the coupling is also not to be overlooked.

As a result, we come to the conclusion that complexity can be constructed as based on rather simple rules: the essential point being that a structure or system may undergo qualitative changes (bifurcations) under the repetitive use of simple rules, or just changing a set of parameters. Said Prigogine—complexity is created through mechanism of bifurcations (in Nash, 1985).

7.3.5. Of main principles

We have seen that complexity can be created by means of simple rules. However, one essential keyword was only mentioned rather than stressed. This is *nonlinearity*,

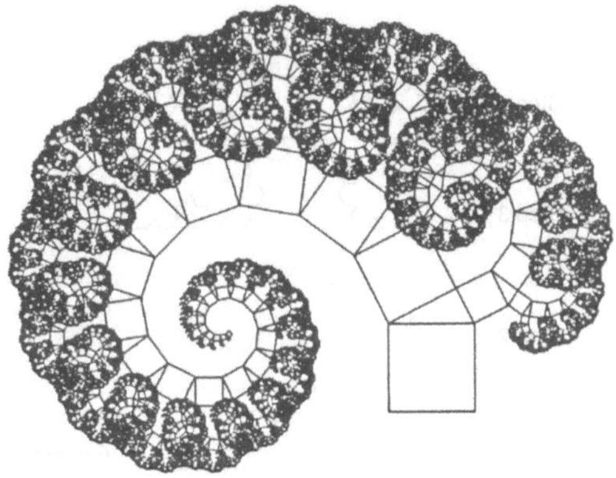

Fig. 7.15. Another variant of the Pythagorean tree after 50 steps (after Peitgen et al., 1992).

a peculiar notion emphasized by a negative prefix. Nonlinearity plays an important role in contemporary science and much has recently been written on the importance of being nonlinear. Actually, the examples shown above (the Mandelbrot set and the quadratic map) involve nonlinearity.

To get rid of the habit of using the property of proportionality and the property of independence (additivity) as basic notions is not so easy because whole generations of scientists have been trained—in the sense of Leonardo da Vinci—to grasp the leading effects in such a convenient way. However, F.Hundertwasser said (in Peitgen and Saupe, 1988): "... the straight line leads to the downfall of mankind. And that line is the rotten foundation of our doomed civilization." Let everyone decide him- or herself whether to agree or not to agree with F.Hundertwasser, but one thing is clear—with linearity there are no large-scale qualitatively new results.

There is a need to distinguish between two opposite routes: from simple to complex and from complex to simple. The first is usually called generalization. Sometimes it is said that mathematical research is largely a process of successive generalization. Starting from basic assumptions (axioms) geometry, rational mechanics and other sciences are built up. The second route is caused by the fact that contemporary science is often able to construct very complicated mathematical models that do not allow direct analyses resp. solutions. Then the only way to understand the process is to find out the main effects. There is certainly a possibility of losing something in the process of simplification. Nevertheless, the history of science knows many examples of how ingenious researchers have found brilliant outcomes. One of such examples is no doubt the Lorenz attractor. The starting point was based on Navier-Stokes equations modelling the circulation and convection of the atmosphere. E.Lorenz has found a simplified version of these equations in the form of a system of nonlinear ordinary differential equations (Lorenz, 1963) (see also Lee, 1992; Stewart, 1989; Thompson and Stewart; 1986). To his surprise, all the richness of chaotic systems was in these equations, resulting in an attractor bearing now his name (Fig. 7.16). In the 3-dimensional phase-space shown in Fig. 7.16, the

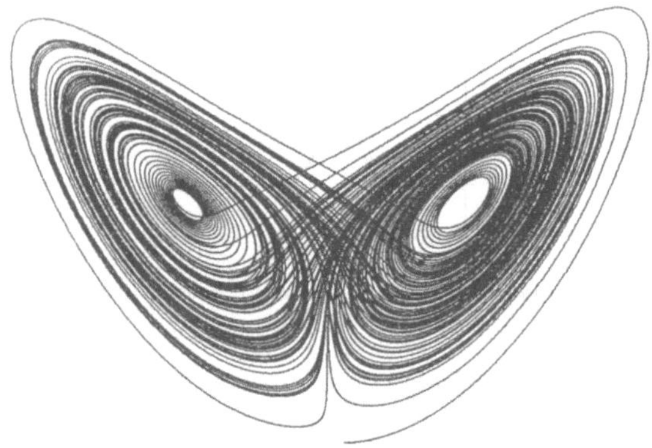

Fig. 7.16. The Lorenz attractor (calculated by V.Miider).

trajectories display a most complex behaviour. Their trace is a kind of a twofold spiral structure that looks like a butterfly with its two wings. It is impossible to predict when a trajectory settled at one wing jumps over to the other wing, etc. That actually explains why the weather prediction becomes impossible under certain conditions. What E.Lorenz actually did, was to discover the skeleton of a complicated system. Everything should be made as simple as possible but not simpler, said Albert Einstein.

Such a route does not always involve a seemingly complex structure like the Lorenz attractor, the outcome from the simplification could also be a complicated theory. In this connection, let us mention solitons. This is again a deep and rich notion of remarkable simplicity (Zabusky and Kruskal, 1965; Drazin and Johnson, 1989). The route to get to it is very similar to that used by Lorenz: from complicated initial equations a model nonlinear evolution equation is derived, the solution to which is the celebrated soliton as a simple structure. Again, Einstein was right—the equation is simple enough but not simpler! Usually a soliton is meant as the Korteweg-de Vries (KdV) soliton, i.e. the solution to the KdV equation that propagates without changing its shape with the amplitude-dependent velocity (Fig. 7.17). In addition, the KdV soliton behaves like a particle when colliding with another soliton. The simplicity of solitons is deep, relating the governing equation to classical eigenvalue problems, and in this sense many facets of classical mechanics become intervowen and get new colour. That is the reason why the soliton concept is so widely used in contemporary hydrodynamics, plasma physics, solid mechanics, quantum mechanics, etc. The theories and methods in solitonics, like the inverse scattering theory, the Painlevé method, the Bäcklund transformation (Drazin and Johnson, 1989; Lamb, 1980; Bishop and Schneider, 1981), are very complicated. The parallel between the concept of soliton and that of heat in the sense of the deepness of theories based upon these notions is obvious.

Let us point out that the Lorenz attractor represents more structural complexity, while solitons form a basis to functional complexity.

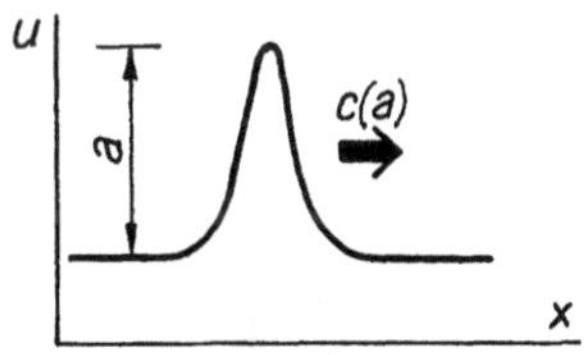

Fig. 7.17. The Korteweg-de Vries soliton.

There are many ways to create complicated systems or structures (see above). Speaking about main principles, we have to point out one, which, at the first glance, may seem naive, but it is a powerful approach—namely, using analogies. Here is a simple example (Stewart, 1989)

- a line segment has 2 end points;
- a square has 4 corners;
- a cube has 8 corners;
- a four-dimensional hypercube has 16 corners;
- a five-dimensional supercube has 32 corners,

and so on. This analogy leads to generalizations in a multidimensional space. The real power of analogies is evident from the Nobel citation in Physics, 1991, stating that P.-G. de Gennes has got the Prize "for discovering that methods developed for studying order phenomena in simple systems can be generalized to more complex forms of matter, in particular to liquid crystals and polymers."

A natural question arises—how to measure complexity. Leaving aside the complexity of theories, which is, without any doubt, a fascinating philosophical problem, let us discuss physically tractable phenomena like solutions, maps, structures, etc., i.e. natural complexity. We may put questions like the following: is fractality of one structure larger than that of another structure, or is one attractor more chaotic than the other? Surprisingly enough there actually are answers to these questions. Fractality is measured by Hausdorff-Besicovich dimension (Mandelbrot, 1977; Kaye, 1989), the attractors are characterized by the spectrum of Lyapunov exponents (Moon, 1987) and chaotic systems by Kolmogorov entropy K (Schuster, 1984). For a regular system K is zero, for a chaotic system K is finite, and for a random (stochastic) system K is infinite. Said I.Stewart: "The brightest ray of light that chaos sheds focuses on the nature of complexity" (1989).

For further research, a working hypothesis can be formulated:

World around us is based on simple but nonlinear rules which are applied repetitively and successively, mostly in time.

That is why we see Nature in its most complex way, and that is why nonlinear science (dynamics) has been so fast developing during the last decades.

7.3.6. Conclusions

What is said above is certainly not a theory of complexity but the ideas about its basic rules as the author sees them (c.f. motto).

The philosophy of science pays a lot of attention to the essence of modern and postmodern (just contemporary) science. According to P.Rosenau (1992), "modern science emphasizes parts rather than wholes, seeking to explain the totality by the sum of the parts..." On the other hand, there are two trends in postmodern science: affirmative and skeptical. Affirmative postmodernists concentrate upon an unbroken wholeness, emphasizing the elegance of complexity in the universe and the richness of difference. Skeptical postmodernists "conceive of the world as fragmented, disrupted, disordered, and in search of instabilities" (Rosenau, 1992). These are actually two facets of the same core—complex Nature based upon holistic principles.

Concentrating upon complexity we may also ask whether it could be a paradigm in contemporary science. The answer is, however, negative—there can be no paradigm of the complex since this is a notion about science not a concept of science (Atlan, 1985).

There is ever so much to be done in understanding complexity. Take for example the human brain with its 10^{11} neurons and 10^{15} synaptic junctions. There is an interesting phenomenon in modelling neural interactions—the usual equality of action and reaction encountered in physics does not hold in the neural network. To build up a fractal system of neural networks and to link it with cognitive behaviour represents an intriguing problem of complexity—perhaps one of the major scientific challenges of the 21st century (Clark, 1989).

The art of asking the right questions is more important than the art of solving them, said Georg Cantor, the inventor of a classical complex structure—the Cantor set. There are many questions about complexity yet to be formulated. The beauty of Nature lies in its complexity often made up of simple things.

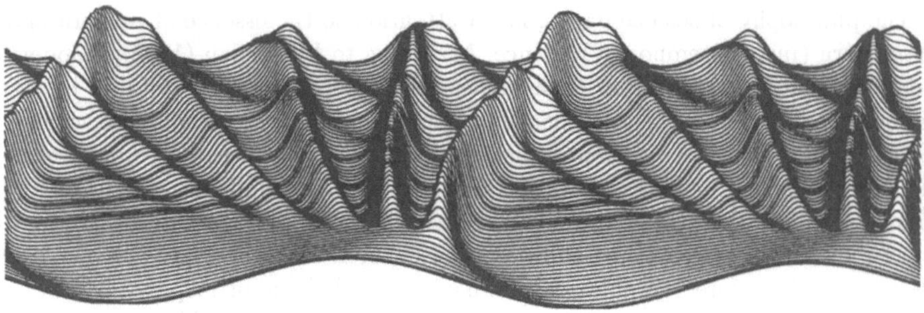

8. Final remarks: complexity of wave motion

The whole is more than the sum of the parts

Aristotle

8.1. THEORIES AND COMPLEXITIES

After presenting the ideas on modelling nonlinear waves, some selected case studies and general concepts of complexity, one should sum up what is said in a more concise form.

It is hoped that the reader will permit a nonscientific prelude: an old story about waves and a man from the far North who was brought up in wild woods and tundras in primitive conditions. Having come to the contemporary world, he got a job as a nightwatch on a lighter carrying bricks. He was left alone for a night and when the crew returned next day, they found the guy throwing bricks overboard just one after another. The raged captain stopped him yelling: "You this and that, why did you do that, etc." The answer was simple: "Look, captain, the bricks have corners but when dropped into water, the waves on water have no corners whatsoever! I only wanted to check if this goes with all the bricks."

Certainly, this peculiar phenomenon could be explained using the language of physics: monopole source, singularities, nearfield, farfield, etc., etc. The story, however, demonstrates the complexity in a straightforward way.

This book is about nonlinear waves. As said in the Preface, we tried to follow a thread from simple basic arguments to complex models and then to simpler governing equations again in order to be able to use the models without loosing the essential features much to the idea of Albert Einstein. Certainly, this is not a novel approach but frequently used by many authors. The story of the physical world is told by people and therefore the personal viewpoint in this book is without any doubt influenced by earlier works. Some of the general ideas on nonlinearities, complexities, waves, etc. have been described in Chapter 7. To sum up what has been said earlier, let us recall some more sayings on theories and complexities because a few sentences here and there stay with us much longer than just a newly-read paper in a good journal.

First, what is to be done? Said Reichenbach (from Dirac, 1980): "It is impossible to know all the causal factors. We can only choose a limited number of significant factors, and use these for predicting future events, being forced to ignore factors having only a minor influence. The hypotheses that certain events are causally connected must be accompanied by a hypothesis about the probable effects of the ignored factors."

Second, what we get then? Said Brown (1987): "Theories, then, often serve to guide observation in at least three ways: they indicate what kind of items exist, what kind of instruments are appropriate for observing them, and how we are to interpret the output from our instruments."

Third, what does it mean? This question is actually answered by Duhem (1954) who considered theory as an economy of thought and as a classification.

Coming to the complexities (see Section 7.3) then we all know that there is nothing more simple than what was discovered yesterday but there is nothing more complicated than what will be discovered tomorrow.

Actually, the word complexity stems from Latin: *com* – together, *plectere* – to plait. It is pretty hard to explain the outcome because the concepts of complexity and simplicity bear the marks of subjectivity (see references in Section 7.3). However, we may state that the external crisscross moving between the Scylla of complexity and the Charybdis of simplicity has enriched our thoughts to better understanding of Nature. Said Lewin (1993): "Complexity means that this is science of qualities not quantities." And M.Gell-Mann has coupled two notions by explaining surface complexity arising out of deep simplicity (see also Lewin, 1993).

However, Truesdell (1984) is not very keen on simplifications, calling mathematical simplicity "a fickle mistress for the student of natural philosophy."

After so many quotations, the author may be blamed relying too much upon them. Alas, "a quotation is a polished prefabricated unit of thought or discourse which has many connotations and associations built in to it" (Mackay, 1988). We have tried to use these "units" for explaining nonlinear waves.

8.2. QUALITIES IN NONLINEAR WAVE MOTION

As said by Maugin (1985): "On certain occasions, one wants to benefit from the nonlinearities. On other occasions, one wants to avoid them." One should only add that nonlinear theory enriches qualitatively our understanding from the world around us. Below the most important effects arising due to nonlinearity, are briefly discussed (Engelbrecht, 1993b)

8.2.1. Distortion of wave profiles (spectral changes)

Due to gradient nonlinearity, the higher harmonics will be generated in the course of propagation. The specific changes certainly depend on the type of nonlinearity (quadratic, cubic, quartic, ...) and on other accompanying effects. The pure nonlinear effect as demonstrated already by Riemann leads to the steepening of the front possibly resulting in the shock formation. Viscosity effects (see Section 5.1.3) are here of the utmost importance. The Voigt-type viscosity leads to a certain Taylor shock profile that has small but finite thickness (Nimmo and Crighton, 1986). The viscosity of the relaxation type leads to smooth solutions for small intensities and to shock waves (discontinuities) for large intensities of the input (Chen, 1973). Similar effects occur also in nonlinear thermoelasticity (see Section 6.2).

The dispersion effects on distortion are even wider. The cubic dispersion may lead to solitary waves with specific spectral changes (see Section 5.3 and the references therein). Additional nonlinearities like the nonlinear driving force may considerably influence the formation of solitary waves (Engelbrecht and Peipman, 1992). If dispersive and viscous effects are both present as in the case of the standard viscoelastic body then the distortion depends very much on the frequency (Jeffrey and Engelbrecht, 1982).

In nonlinear thermoelasticity with nonlinear equation of that conduction (see Section 3.2), the temperature localization contrary to the usual smoothening effects can occur (Kudryumov et al., 1988).

8.2.2. Amplitude-dependent velocities

In fact, shock waves and smooth constant profile waves are possible only because of nonlinearities. Their formation occurs because the various parts of a wave profile propagate with different velocities resulting in a specific profile and a specific velocity. This is explicitly the case of the celebrated soliton—dispersion only will lead to frequency dependent group velocity (Chapter 1) but the same effect together with the nonlinearity results in the amplitude dependent soliton velocity (cf. Section 5.3).

The following list gives the specific velocities in nonlinear wave motion:
c_0 – sound velocity:
c_{sh} – shock velocity;
c_{sol} – soliton velocity;
c_e – equilibrium velocity;
c_i – instantaneous velocity;
c_{st} – steady state velocity.

In addition, other interesting phenomena can occur in nonlinear systems. For example, anisochroism may happen in resonators that means a relative change in the velocity of the fundamental frequency as a result of nonlinearities (Maugin, 1985). The velocity of surface waves may be affected by nonlinear dispersion (Parker and Talbot, 1985). The list of these examples is by no means complete.

8.2.3. Interaction of waves

This may lead to the coupling of various effects. For example, the nonlinear coupling of 2D transverse and longitudinal waves leads to a certain skew distortion of transverse waves (Peipman et al., 1992) while longitudinal waves are not affected. In the case of surface waves, convolution may occur in devices called acoustic convolvers where the interaction components at the summed frequency are generated (Maugin,

1985). The nonlinear reflection from a boundary will cause phase shifts resulting in possible narrowing or broadening of the pulse (Engelbrecht, 1983). The reason is in the interaction of incident and reflecting waves and certainly the transmitted wave is also affected (Litvin and Tsvankin, 1987). As a matter of fact, the coupling of deformation with other fields leads also to many new effects (Maugin et al., 1986). In the solidification of dilute binary mixtures, nonlinear couplings among heat, mass, and momentum transfer cause strong nonlinear terms like $\nabla(u\nabla u)$ and $\nabla(\nabla^2 u\nabla u)$ in the governing equations (Riley, 1990). The modulational instability may occur in a micropolar medium if two circularly polarized transverse waves are nonlinearly coupled (Erbay et al., 1992).

8.2.4. Applications

We can really benefit from nonlinear phenomena in many areas, like
- nonlinear solid mechanics;
- nonlinear acoustics;
- nonlinear seismology;
- nonlinear nondestructive testing (NDT);
- nonlinear surface acoustic waves (SAW) devices;

and many others.

The benefits are related to the richer description of physical effects that leads directly to the better usage of materials (economy) and the wider knowledge about the process (information). The latter is especially important because the process itself does not depend on our description (linear or nonlinear) but proceeds in its own way. Wider information may considerably enrich the quality of solving the inverse problems, for example in the NDT. Besides the benefits we would also like to get rid of some effects if possible, like shock waves causing stress concentration, the loss of harmonicity, unwanted effects caused by nonlinear coupling, etc.

A good example how to convert a shock wave into a smooth one by restructuring the waveguide is given by Sugimoto (1993) — see also Section 7.2. This way or another, the world is much richer than its linear approximation.

8.3. QUESTIONS AND PROSPECTS

The importance of being nonlinear is nowadays widely accepted in many branches of science. In practical applications the taming of nonlinear systems has been slower that is quite understandable due to the mathematical difficulties. Many problems in high-technology areas, however, need nonlinearities to be taken into account in order to guarantee qualitatively correct results. There are many hot problems and the following list is by no means exhaustive:

(i) nonlinear continuum mechanics: description of material inhomogeneities (Maugin, 1993), observable and internal variables, etc.;

(ii) nonlinear micromechanics with possible accounting for structural features such as phase distribution, microcracks, porosity; damage (Barenblatt, 1993); smart materials;

(iii) nonlinear coupling: nonlinear theories of magnetoelasticity, electroelasticity (for ceramics), thermoelasticity (for large temperature gradients) and possible thermal "solitons";
(iv) coupling of nonlinearity and other "non-properties" (Maugin et al., 1992), like nonquadratic nonconvex potentials, anisotropy, nonlocality, symmetry breaking, etc.;
(v) nonlinear waves in structures with attention to boundaries and geometrical constraints: propagating instabilities (Kyriakides, 1993), waves in waveguides (Parker, 1994);
(vi) solitonics: high-order dispersion, weakly and strongly nonconservative systems, energy influx;
(vii) quest for special types of waves: exceptional waves (Abbudi and Garnett, 1990), crenellated fronts (Caroli et al., 1990), etc.;
(viii) spatio-temporal chaos and predictability of wave motion.

8.4. ENDNOTE

From the viewpoint of mechanics, nonlinear wave motion is a relatively new area. Mechanics, as originally understood by Greeks ($\mu\eta\chi\breve{\alpha}\nu\acute{\eta}$) was related to a device for delusion in warfare. Much has been initiated by mankind in this relation. Although such a driving force has not disappeared in our fragile world, contemporary understanding of mechanics makes use of old traditions together with new ideas and the outcome is meant for better mankind. Nonlinear wave motion means complexities and complexities mean the hard work to grasp essentials. When we reach it then we understand that "the inner nature of the cosmos is a landscape of exquisite beauty which appropriately underlies the exterior splendour of our experience" (Oliver, 1991).

Two statements are needed finally. The first is:

Scratch mechanics and you find nonlinearity.

The second is based on words of Thomas Aquinas (see MacKay, 1988): "Practical sciences proceed by building up, theoretical sciences by resolving into components". Consequently we may state:

Mechanics is very much both theoretical and practical.

The nonlinear wave dynamics is full of theoretical surprises and practical ideas, it is full of complexities and deep simplicities. This book is about them. Paul Dirac would have asked why it took so many pages to say it. The author leaves the decision about the length and the contents of this treatise to the reader.

References

Abbudi, M. and Barnett, D.M. (1990) Search for Exceptional Waves: Tellurium Dioxide, an Example, *Appl. Mech. Rev.*, **43**(5), Part 2, 354–358.

Abe, K. and Abe, T. (1979) Recurrence of Initial State of the Korteweg-de Vries Equation, *Phys. Fluids*, **22**, 1644–1646.

Abe, K. and Satofuka, N. (1981) Recurrence of Initial State of Nonlinear Ion Waves, *Phys. Fluids*, **24**, 1045–1048.

Ablowitz, M.J. and Clarkson, P.A. (1991) *Solitons, Nonlinear Evolution Equations and Inverse Scattering*, Cambridge University Press.

Atlan, H. et al. (1985) *La Sfida della Complessita*, A. Cura di Gianluca. Bocchie Mauro Ceruti, Feltrinelli, Milano.

Bachelard, G. (1934) *Le Nouvel Esprit Scientifique*, Presses Universitaires de France, Paris.

Bachelard, G. (1940) *La Philosophie du Non*, Presses Universitaires de France, Paris.

Bacon, D.R. (1986) Non-Linear Ultrasonic Fields: Theory and Experiment, *Proc. IOA*, **8**(2), 39–46.

Bamman, D.J. (1985) An Internal Variable Model of Elastoviscoplasticity, in E.C.Aifantis and J.P.Hirth (Eds.), *The Mechanics of Dislocations*, Amer. Soc. of Metals, Metals Park, Ohio, 203–221.

Barenblatt, G.I. (1993) Micromechanics of Fracture, in S.R.Bodner, J.Singer, A.Solan, Z.Hashin (Eds.), *Theoret. and Appl. Mech. 1992*, Elsevier, New York, 25–52.

Bataille, K. and Lund, F. (1982) Nonlinear Waves in Elastic Media, *Physica*, **60**, 95–104.

Bhatnagar, P.L. (1979) *Nonlinear Waves in One-Dimensional Dispersive Systems*, Clarendon Press, Oxford.

Bell, J.E. (1973) The Experimental Foundation of Solid Mechanics, in S.Flügge (Ed.), *Enc. of Phys.*, **VIa/1**, Springer, Berlin et al., 1–813.

Bishop, A.R. and Schneider, T. Eds. (1978) *Soliton and Condensed Matter Physics*, Springer, Berlin.

Bishop, A.R., Krumhansl, J.A. and Trullinger, S.E. (1980) Solitons in Condensed Matter: a Paradigm, *Physica*, **D1**, 1–44.

Bjørnø, I. (1986) Characterization of Biological Media by Means of their Non-Linearity, *Ultrasonics*, **24**, 254–259.

Bland, D.R. (1969) *Nonlinear Dynamic Elasticity*, Blaisdell, Waltham, Mass. et al.

Bland, D.R. (1988) *Wave Theory and Applications*, Clarendon Press, Oxford.

Bogdanov, A.N. and Skvortsov, A.T. (1992) Nonlincar Elastic Wavcs in a Granular Medium, *J. Physique* IV, **2**, 1779–1782.

Boussinesq, M.J. (1871a) Théorie Générale des Mouvements Qui Sont Propagés dans un Canal Rectangulaire Horizontal, *C.R. Acad. Sci. Paris*, **73**, 256–260.

Boussinesq, M.J. (1871b) Théorie de l'Intumescence Appelée Onde Solitaire ou de Translation se Propageant dans un Canal Rectangulaire, *C.R. Acad. Sci. Paris*, **72**, 755–759.

Boyd, J.P. (1990) New Directions in Solitons and Nonlinear Periodic Waves: Polycnoidal Waves, Imbricated Solitons, Weakly Nonlocal Solitary Waves and Numerical Boundary Value Algorithms, *Adv. in Appl. Mech.*, **27**, 1–82.

Bracewell, R.N. (1978) *The Fast Fourier Transform and its Applications*, McGraw-Hill, New York.

Brankov, G. (1978) *Basic Biomechanics*, Bulgarian Acad. Sci., Sofia (in Bulgarian).

Braun, M. and Engelbrecht, J. (1995) *Nonlinear Waves in Nonlocal Media*, Res. Rept. Mech., 136/95, Inst. of Cybernetics, Estonian Acad. Sci., Tallinn.

Breazeale, M.A. and Philip, J. (1984) Determination of Third Order Elastic Constants from Ultrasonic Second Harmonic Generation Measurements, in W.P.Mason and R.N.Thurstov (Eds.), *Phys. Acoustics*, **XVII**, Academic Press, New York, 1–60.

Brown, H.I. (1987) *Observation and Objectivity*, Oxford University Press, New York and Oxford.

Bullough, R.K. and Caudrey, P.I. Eds. (1980) *Solitons*, Springer, Berlin.

Bunde, A. and Havlin, S. Eds. (1991) *Fractals and Disordered System*, Springer, Berlin et al.

Cantrell, J.H. (1989) Acoustic Nonlinearity Parameters and Higher-Order Elastic Constants of Crystals, *Proc. IOA*, **11**, Part 5, 445–452.

Caroli, B., Caroli, C. and Roulet, B. (1990) Directional Solidification of a Faceted Crystal, *Appl. Mech. Rev.*, **43**(5), Part 2, 53.

Cartwright, N. (1984) *How the Laws of Physics Lie*, Clarendon Press, Oxford.

Casti, J.L. and Karlqvist, A. Eds. (1986) *Complexity, Language and Life: Mathematical Approaches*, Springer, Berlin.

Chambers's Twentieth Century Dictionary (1961) Chambers, Edinburgh et al.

Chambers Dictionary of Science and Technology (1982) Chambers, Edinburgh.

Chandrasekharaiah, D.S. (1986) Thermoelasticity with Second Sound: a Review, *Appl. Mech. Rev.*, **39**, 355–376.

Chandrasekharaiah, D.S. and Debnath, L. (1994) *Continuum Mechanics*, Acad. Press, Boston, Mass.

Chen, P.J. (1973) Growth and Decay of Waves in Solids, in S.Flügge (Ed.), *Enc. of Phys.*, **VIa/3**, Springer, Berlin et al.

Chivers, R.C. (1981) Tissue Characterization, *Ultrasound Med. Biol.*, **7**, 1–20.

Chivers, R.C. and Hill, C.R. (1975) A Spectral Approach to Ultrasonic Scattering from Tissue: Methods, Objectives and Backscattering Measurements, *Phys. Med. Biol.*, **20**, 799–815.

Christensen, R.M. (1971) *Theory of Viscoelasticity*, Academic Press, New York.

Clark, J.W. (1989) Introduction to Neural Networks, in A.N.Proto (Ed.), *Nonlinear Phenomena in Complex Systems*, North-Holland, Amsterdam et al., 1–102.

Clifton, R.J. (1993) Analysis of Failure Waves in Glasses, *Appl. Mech. Rev.*, **46**,

540–546.

Cole, J.D. (1968) *Perturbation Methods in Applied Mathematics*, Blaisdell, Waltham, Mass.

Cole, S.C. (1972) *Membranes, Ions and Impulses. A Chapter of Classical Biophysics*, Univ. of California Press, Berkeley et al.

Coleman, B.D. and Noll, W. (1961) Foundations of Linear Viscoelasticity, *Rev. Mod. Phys.*, **33**, 239–249.

Colli, P. (1989) On a Nonlinear and Nonlocal Evolution Equation Related to Muscle Contraction, *Nonlinear Analysis, Theory, Methods and Applications*, **13**(10), 1149–1162.

Collins English Dictionary (1989) Collins, London and Glasgow.

Courant, R. and Hilbert, D. (1963) *Methods of Mathematical Physics, II*, J.Wiley-Interscience, New York.

Crandall, S.H. (1974) Nonlinearities in Structural Dynamics, *Shock and Vibration Digest*, **6**(8), 1–13.

Crum, L.A., Daniels, S., Dyson, M., ter Haar, G.R. and Walton, A.J. (1986) Acoustic Cavitation and Medical Ultrasound, *Proc. IOA*, **8**(2), 137–146.

David, E.A. and Parker, D.F. (1988) Nonlinear Evolution of Piezoelectric SAWs, in D.F.Parker and G.A.Maugin (Eds.), *Recent Developments in Surface Acoustic Waves*, Springer, Berlin, 21–29.

Davies, M.A. and Moon, F.C. (1993) 3-D Spatial Chaos in the Elastica and the Spinning Top: Kirchhoff Analogy, *Chaos*, **3**(1), 93–99.

Davydov, A.S. (1985) *Solitons in Molecular Systems*, Reidel, Dordrecht et al.

Dictionary of Scientific Bibliography (1973) Charles Scribner's Sons, New York, Vol. VII.

Dirac, P.A. (1980) Why We Believe in the Einstein Theory, in B.Gruber and R.S.Millmann (Eds.), *Symmetrics in Science*, Plenium Press, New York and London, 1–11.

Dodd, R.K., Eilbeck, J.C., Gibbon, J.D. and Morris, H.C. (1982) *Solitons and Nonlinear Wave Equations*, Academic Press, London et al.

Drazin, P.G. and Johnson, R.S. (1989) *Solitons: an Introduction*, Cambridge Univ. Press, Cambridge.

Duhem, P. (1954) *The Aim and Structure of Physical Theory*, Princeton University Press (translated from the 2nd French ed. 1914).

Edelen, D.G.B. (1976) Nonlocal Field Theories, in A.C.Eringen (Ed.), *Continuum Physics, IV—Polar and Nonlocal Field Theories*, Acad. Press, New York et al., 75–204.

Efinger, H.J. (1984) On the Theory of Certain Nonlinear Schrödinger Equations with Nonlocal Interaction, *Il Nuovo Cimento*, **80B**(2), 260–278.

Eilenberger, G. (1981) *Solitons, Mathematical Methods for Physicists*, Springer, Berlin et al.

Eleonsky, V.M., Korolev, V.G., Kulagin, N.E. and Shilnikov, L.P. (1994) Dynamical Systems in the Theory of Nonlinear Waves with Allowance for Nonlocal Interactions (the Whitham-Benjamin Equation), *Chaos*, **4**(2), 377–384.

Encyclopedia Britannica (1769) Bell and Macfarquhar, Edinburgh, Vol. 1.

Encyclopedia Britannica (1964) Benton, London et al., Vol. 3.

Engelbrecht, J. (1981) On Theory of Pulse Transmission in a Nerve Fibre, *Proc. Roy. Soc. London*, **A375**, 195–209.

Engelbrecht, J. (1983) *Nonlinear Wave Processes of Deformation in Solids*, Pitman, London.

Engelbrecht, J. Ed. (1989) *Nonlinear Waves in Active Media*, Springer, Heidelberg et al.

Engelbrecht, J. (1990) Ordered Definitions in the Theory of Nonlinear Waves, *Proc. Estonian Acad. Sci., Phys. Math.*, **39**(3), 252–257.

Engelbrecht, J. (1991) *An Introduction to Asymmetric Solitary Waves*, Longman, Harlow.

Engelbrecht, J. (1992) Etalon Evolution Equations, *Proc. Estonian Acad. Sci., Phys. Math.*, **41**(2), 77–83.

Engelbrecht, J. (1993a) Complexity and Simplicity, *Proc. Estonian Acad. Sci., Phys. Math.*, **42**(1), 107–118.

Engelbrecht, J. (1993b) Qualitative Aspects of Nonlinear Wave Motion: Complexity and Simplicity, *Appl. Mech. Rev.*, **46**(12), Part 1, 509–518.

Engelbrecht, J. (1995a) Beautiful Dynamics, *Proc. Estonian Acad. Sci., Phys. Math.*, **44**(1), 108–119.

Engelbrecht, J. (1995b) Waves, Solids, and Nonlinearities, *Shock and Vibr.*, **2**(2), 173–190.

Engelbrecht, J.K., Fridman, V.E. and Pelinovski, E.N. (1988) *Nonlinear Evolution Equations*, Longman, Harlow.

Engelbrecht, J. and Khamidullin, Y. (1988) On the Possible Amplification of Nonlinear Seismic Waves, *PEPI*, **50**, 39–45.

Engelbrecht, J. and Chivers, R.C. (1989) Evolution Equations and Ultrasonic Wave Propagation in Biological Tissues, *Phys. Med. Biol.*, **34**, 1571–1592.

Engelbrecht, J. and Peipman, T. (1992) Nonlinear Waves in a Layer with Energy Influx, *Wave Motion*, **16**, 173–181.

Engelbrecht, J. and Ravasoo, A. (1994) From Continuum Mechanics to Applications in Nondestructive Testing, *Bull. Techn. Univ. Istanbul*, **47**(1–2), 83–103.

Engelbrecht, J., Salupere, A., Peipman, T. and Maugin, G.A. (1995) Spectral Analysis of Soliton Formation, in A.Mielke and K.Kirchgässner (Eds.), *Proc. IUTAM/ISIMM Symp. on Structure and Dynamics of Nonlinear Waves in Fluids*, World Sci., Singapore, 216–222.

Engelbrecht, J. and Maugin, G.A. (1996) Deformation Waves in Thermoelastic Media and the Concept of Internal Variables, *Arch. Appl. Mech.*, **66**, 200–207.

Epstein, M. (1993) The Long Road from Uniformity to Defects, *Materials Science Forum*, **123–125**, 427–434.

Erbay, S., Erbay, H.A. and Dost, S. (1992) Nonlinear Wave Interactions in a Micropolar Elastic Medium, *Wave Motion*, **16**, 163–172.

Eringen, A.C. (1962) *Nonlinear Theory of Continuous Media*, McGraw-Hill, New York et al.

Eringen, A.C. (1992) Vistas of Nonlocal Continuum Physics, *Int. J. of Eng. Science*, **30**(10), 1551–1565.

Eringen, A.C. and Kafadar, C.B. (1976) Polar Field Theories, in A.C.Eringen (Ed.), *Continuum Physics, IV—Polar and Nonlocal Field Theories*, Acad. Press, New York et al., 1–73.

Eringen, A.C. and Maugin, G.A. (1990) *Electrodynamics of Continua, I, II*, Springer, New York et al.

Erofeyev, V.I. and Potapov, A.I. (1993) Longitudinal Strain Waves in Nonlinearly-Elastic Media with Couple Stresses, *J. of Non-Linear Mech.*, **28**(4), 483–488.

Eshelby, J.D. (1951) The Force on an Elastic Singularity, *Phil. Trans. Roy. Soc. London*, **244**, 87–112.

Ewing, W.M., Jardetzky, W.S. and Press, F. (1957) *Elastic Waves in Layered Media*, McGraw-Hill, New York.

Feigenbaum, M.J. (1980) Universal Behaviour in Nonlinear Systems, *Los Alamos Science*, **1**, 4–27.

Feldmann, M., Chivers, R.C. and Engelbrecht, J. (1990) On the Use of Evolution Equations for Describing the Nearfield Zone of Ultrasonic Transducers, *Acoust. Lett.*, **13**(8), 130–135.

Fermi, A., Pasta, J. and Ulam, S. (1955) *Studies of Nonlinear Problems, I*, Los Alamos Report (reproduced in *Lect. Appl. Math.*, 1974, **15**, 143–156).

Ferry, J.D. (1961) *Viscoelastic Properties of Polymers*, J.Wiley, New York and London.

Fischer, R.A. (1937) The Wave of Advance of Advantageous Genes, *Ann. Eugenics*, **7**, 355–369.

Flügge, W. (1967) *Viscoelasticity*, Blaisdell, Waltham, Mass. et al.

Fokas, A.S. and Zakharov, V.E. Eds. (1993) *Important Developments in Soliton Theory*, Springer, Berlin et al.

Fornberg, B. and Whitham, G.B. (1978) A Numerical and Theoretical Study of Certain Nonlinear Wave Phenomena, *Phil. Trans. Roy. Soc. London*, **A289**, 373–404.

Fordy, A.P. Ed. (1990) *Soliton Theory: a Survey of Results*, Manchester University Press.

Di Francesco, D. and Noble, D.A. (1985) A Model of Cardiac Electrical Activity Incorporating Ionic Pumps and Concentration Changes, *Phil. Trans. Roy. Soc. London*, **B307**, 353–398.

Francis, P.H. (1972) Thermo-Mechanical Effects in Elastic Wave Propagation: a Survey, *J. Sound Vibr.*, **21**(2), 181–192.

Gardner, C.S., Green, J.M., Kruskal, M.D. and Miura, R.M. (1967) Method for Solving the Korteweg-de Vries Equation, *Phys. Rev. Lett.*, **19**, 1095–1097.

Cleick, J. (1988) *Chaos. Making a New Science*, Heinemann, London.

Gorschkov, K.A., Ostrovski, L.A. and Pelinovski, E.N. (1974) Some Problems of Asymptotic Theory of Nonlinear Waves, *Proc IEEE*, **62**, 1511–1517.

Green, A.E. and Rivlin, R.S. (1964) Multipolar Continuum Mechanics, *Arch. for Rat. Mech. and Anal.*, **17**, 113–147.

Grindrod, P. (1991) *Patterns and Waves*, Clarendon Press, Oxford.

Gusev, A.A. (1988) Two Dilatancy-Based Models to Explain Coda-Wave Precursors and P/S Spectral Ratio, *Tectonophys.*, **152**, 227-237.

Guz, A.N. (1986) *Elastic Waves in Prestressed Solids, I, II*, Naukova Dumka, Kiev (in Russian).

Gödel, K. (1931) Über Formal Unentscheidbare Sätze der Principia Mathematica und Verwandter Systeme, *Monatshefte für Math. und Physik*, **38**, 173–198.

Hamilton, M.F., Il'insky, Yu.A. and Zabolotskaya, E.A. (1995a) Local and Nonlocal Nonlinearity in Rayleigh Waves, *J. Acoust. Soc. Am.*, **97**(2), 882–890.

Hamilton, M.F., Il'insky, Yu.A. and Zabolotskaya, E.A. (1995b) Evolution Equations for Nonlinear Rayleigh Waves, *J. Acoust. Soc. Am.*, **97**(2), 891–897.

Han, C.D. (1976) *Rheology in Polymer Processing*, Academic Press, New York et al.

Hansen, P.J. and Lonngren, K.E. (1993) On the Prediction of the Number of Solitons Excited by an Arbitrary Potential: an Observation from Inverse Scattering, *Physica*, **D68**, 12–17.

Hodgkin, A.L. and Huxley, A.F. (1952) Quantitative Description of Membrane Current and its Application to Conduction and Excitation in Nerve, *J. Physiol.*, **117**, 500–544.

Hovis, R.C. and Kragh, H. (1993) P.A.M.Dirac and the Beauty of Physics, *Sci. Amer.*, **268**(5), 104–109.

Humprey, V.E., Burgess, M. and Sampson, N. (1986) Harmonic Generation Due to Non-Linear Propagation in a Focussed Ultrasonic Field, *Proc. IOA*, **8**(2), 47–54.

Jeffrey, A. (1994) Exact and Asymptotic Methods in Nonlinear Wave Theory, in A.Jeffrey and J.Engelbrecht (Eds.), *Nonlinear Waves in Solids*, Springer, Wien and New York, 1–50.

Jeffrey, A. and Kakutani, T. (1972) Weak Nonlinear Dispersive Waves, *SIAM Rev.*, **14**, 582–643.

Jeffrey, A. and Engelbrecht, J. (1982) Waves in Nonlinear Relaxing Media, in F.Mainardi (Ed.), *Wave Propagation in Viscoelastic Media*, Pitman, Boston et al., 95–123.

Jeffrey, A. and Engelbrecht, J. Eds. (1994) *Nonlinear Waves in Solids*, Springer, Wien et al.

Johnson, S. (1755) *Dictionary of the English Language*, Vol. 1, W.Strahan, London.

Johnson, A.F. (1975) Pulse Propagation in Heat-Conducting Elastic Materials, *J. Mech. Phys. Solids*, **23**, 55–67.

Kadomtsev, B.B. (1988) *Collective Phenomena in Plasma*, Nauka, Moscow, (in Russian).

Kalda, J. (1995) Periodic Solutions of the Korteweg-de Vries Equation and the Number of Eigenvalues, *Proc. Estonian Acad. Sci., Phys. Math.*, **44**(1), 88–95.

Kampus, G. and Csányi, V.J. (1987) Notes on Order and Complexity, *Theor. Biol.*, **124**, 111–121.

Kawahara, T. and Toh, S. (1988) Pulse Interaction in an Unstable Dissipative-Dispersive Nonlinear System, *Phys. of Fluids*, **31**(8), 2103–2111.

Kaye, B.H. (1989) *A Random Walk through Fractal Dimensions*, VCH, Weinheim.

McKean, H.P. (1970) Nagumo's Equation, *Adv. in Math.*, **4**, 209–223.

Kolmogorov, A., Petrovsky, I. and Piscounoff, N. (1937) Etude des Equations de la Diffusion Avec Croissance de la Quantité de Matière; Son Application à un Problème Biologique, *Bull. Univ. Moscou*, **1**, 1–25.

Kolsky, H. (1963) *Stress Waves in Solids*, 2nd ed., Dover, New York.

Konno, K., Ichikawa, Y.H. and Wadati, M. (1981) A Loop Soliton Propagating along a Stretched Rope, *J. Phys. Soc. Japan*, **50**, 1025–1026.

Konno, K. and Jeffrey, A. (1994) The Loop Soliton, in L.Debnath (Ed.), *Advances in Nonlinear Waves*, Pitman, London, 162–182.

Korteweg, D.J. and de Vries, G. (1895) On the Change of Form of Long Waves Advancing in a Rectangular Channel and on a New Type of Long Stationary Waves, *Philos. Mag.*, Ser. 5, **39**, 422–443.

Kosevich, A.M. (1988) *Theory of Crystal Lattices*, Vyshscha Shkola, Kiev (in Russian).

Kozak, J. and Šilený, J. (1985) Seismic Events with Non-Shear Component: I Shallow Earthquakes with a Possible Tensile Source Component, *PAGEOPH*, **123**, 1–15.

Krautkramer, J. and Krautkramer, H. (1977) *Ultrasonic Testing of Materials*, Springer, Berlin.

Kreiss, H.O. and Oliger, J. (1972) Comparison of Accurate Methods for the Integration of Hyperbolic Equation, *Tellus*, **24**, 199–215.

Kröner, E. (1968) Interrelations between Various Branches of Continuum Mechanics, in E.Kröner (Ed.), *Mechanics of Generalized Continua*, Springer, Berlin et al., 330–340.

Kröner, E. (1993) Configurational and Material Forces in the Theory of Defects in Ordered Structures, *Materials Science Forum*, **123–125**, 447-454.

Kuc, R.C. (1980) Clinical Application of an Ultrasound Attenuation Estimation Technique for Liver Pathology Characterization, *IEEE, Trans. Biomed. Eng.*, **BME 27**, 313-319.

Kudryumov, S.P., Malinetski, G.G., Potapov, A.B. and Samarski, A.A. (1988) Structures in Nonlinear Media, in A.A.Samarski (Ed.), *Computers and Nonlinear Phenomena*, Nauka, Moscow, 5–43 (in Russian).

Kuhn, T. (1977) *The Essential Tension*, The University of Chicago Press, Chicago and London.

Kunin, I.A. (1982) *Elastic Media with Microstructure, I: One-Dimensional Models*, Springer, Berlin et al.

Kunin, I.A. (1983) *Elastic Media with Microstructure, II: Three-Dimensional Models*, Springer, Berlin et al.

Kyriakides, S. (1993) Propagating Instabilities in Structures, *Adv. in Appl. Mech.*, **30**, 67–189.

Lamb, H. (1906) *Hydrodynamics*, Cambridge University Press, 3rd ed.

Lamb, G.L. Jr. (1980) *Elements of Soliton Theory*, J.Wiley, New York et al.

Landau, L.D. (1941) The Theory of Superfluidity of Helium II, *J. Phys. USSR*, **5**, 71–90.

Lardner, R.W. (1969) The Linear Theory of Second-Grade Elastic Materials, *Quarterly of Appl. Math.*, **27**, 323–334.

Lebon, G., Iou, D. and Casas-Vazquez, J. (1992) Questions and Answers about a Thermodynamic Theory of the Third Type, *Contemporary Phys.*, **33**, 41–51.

Lee, P.C.Y. (1976) Some Problems in Vibrations of Piezoelectric Crystal Plates, in

J.D.Achenbach, Y.H.Pao and H.F.Tiersten (Eds.), *Applications of Elastic Waves in Electrical Devices, Non-Destructive Testing and Seismology*, Northwestern University Press, Chicago, 442–493.

Lee, R. (1992) Readings in the "New Science", *Review (Fernand Braudel Center)*, **15**, 113–171.

Lewin, R. (1993) *Complexity. Life at the Edge of Chaos*, J.M.Dent Ltd., London.

Lieberstein, H.M. (1967) On the Hodgkin-Huxley Partial Differential Equation, *Math. Biosci.*, **1**, 45–69.

Lighthill, J. (1956) Viscosity Effects in Sound Waves of Finite Amplitude, in G.K.Batchelor and R.M.Davies (Eds.), *Surveys in Mechanics*, University Press, Cambridge, 255–351.

Lighthill, J. (1986) The Recently Recognized Failure of Predictability in Newtonian Dynamics, *Proc. Roy. Soc. London*, **A407**, 35–50.

Litvin, A.L. and Tsvankin, I.D. (1987) Interaction of Plane Waves with the Boundary of a Nonlinear Elastic Medium, in A.N.Nikolaev and I.N.Galkin (Eds.), *Problems of Nonlinear Seismology*, Nauka, Moscow, 128–136 (in Russian).

Lonngren, K. and Scott, A., Eds, (1978) *Solitons in Action*, Academic Press, New York et al.

Lord, H.V. and Shulman, Y. (1967) A Generalized Dynamical Theory of Thermoelasticity, *J. Mech. Phys. Solids*, **15**, 299–309.

Lorenz, E.N. (1963) Deterministic Nonperiodic Flow, *J. Atmos. Sci*, **20**, 130–141.

Love, A.E.N. (1906) *A Treatise on the Mathematical Theory of Elasticity*, Cambridge University Press.

Lund, F. (1983) Interpretation of the Precursor of the 1960 Great Chilean Earthquake as a Seismic Solitary Wave, *PAGEOPH*, **121**, 17–26.

Mackay, A.L. (1988) *The Harvest of a Quiet Eye*, The Institute of Physics, Bristol and London, Crane, Russak & Co, Inc., New York.

Mandel, J. (1980) Variables Cachées, Puissance Dissipée, Dissipativité Normal, *Sci. et Techn. de l'Armement, Special Issue (Jan.)*, 37–39.

Mandelbrot, B.B. (1977) *Fractals: Form, Chance and Dimension*, Freeman, San Francisco.

Mandelbrot B.B. (1982) *The Fractal Geometry of Nature*, Freeman, New York.

Manton, N. (1994) *Soliton Dynamics*, Seminar at Isaac Newton Institute, Cambridge, Jan. 31st. (oral information).

Maslov, V.P. and Mosolov, P.P. (1981) Oscillations of Bimodular Rods, *Soviet Math.-Uspekhi*, **36**, 240–241.

Maugin, G.A. (1979) Nonlocal Theories of Gradient-Type Theories: A Matter of Convenience?, *Archives of Mechanics (Archiwum Mechaniki Stosowanej)*, **31**(1), 15–26.

Maugin, G.A. (1985) *Nonlinear Electromechanical Effects and Applications*, World Scientific, Singapore.

Maugin, G.A. (1990) Internal Variables and Dissipative Structures, *J. Non-Equilib. Thermodyn.*, **15**, 173–192.

Maugin, G.A. (1992) Material Forces in Homogeneous Elastoplastic Solids, in W.Muschik and G.A.Maugin (Eds.), *Nonlinear Thermodynamical Processes in*

Continua, TU Berlin, TUB Documentation, Heft **61**, 175–186.

Maugin, G.A. (1992a) Applications of an Energy-Momentum Tensor in Nonlinear Elastodynamics: Pseudomomentum and Eshelby Stress in Solitonic Elastic Systems, *J. Mech. Phys. Solids*, **40**, 1543–1558.

Maugin, G.A. (1992b) Nonlinearity for Nonscientists: Bridging the Two Cultures?, in W.Muschik and G.A.Maugin (Eds.), *Nonlinear Thermodynamical Processes in Continua*, TU Berlin, TUB Documentation, Heft **61**, 217–245.

Maugin, G.A. (1993) *Material Inhomogeneities in Elasticity*, Chapman and Hall, London.

Maugin, G.A. (1994a) *The Thermomechanics of Plasticity and Fracture*, Cambridge University Press, Cambridge, Mass.

Maugin, G.A. (1994b) Physical and Mathematical Models of Nonlinear Waves in Solids, in A.Jeffrey and J.Engelbrecht (Eds.), *Nonlinear Waves in Solids*, Springer, Wien et al., 109–233.

Maugin, G.A. (1995) On Some Generalization of Boussinesq and KdV Systems, *Proc. Estonian Acad. Sci., Phys. Math.*, **44**(1), 40–55.

Maugin, G.A., Collet, B. and Pouget, J. (1986) Nonlinear Wave Propagation in Coupled Electromechanical Systems, in T.W.Wright (Ed.), *Nonlinear Wave Propagation in Mechanics*, ASME, New York, 57–84.

Maugin, G.A. and Cadet, S. (1991) Existence of Solitary Waves in Martenistic Alloys, *Int. J. of Eng. Sci.*, **29**, 243–255.

Maugin, G.A. and Hadouaj, H. (1991) Solitary Surface Transverse Waves on an Elastic Substrate Coated with a Thin Film, *Phys. Rev.*, **B44**, 1266–1280.

Maugin, G.A., Collet, B., Drouot, R. and Pouget, J. (1992) *Nonlinear Electromechanical Couplings*, J.Wiley, Chichester et al.

Maugin, G.A. and Muschik, W. (1994) Thermodynamics with Internal Variables, I—General Concepts, *J. of Non-Equilib. Thermodyn.*, **19**, 217–249.

Maugin, G.A. and Muschik, W. (1994) Thermodynamics with Internal Variables, II—Applications, *J. of Non-Equilib. Thermodyn.*, **19**, 250–289.

Maugin, G.A. and Engelbrecht, J. (1994) A Thermodynamical Viewpoint on Nerve Pulse Dynamics, *J. of Non-Equilib. Thermodyn.*, **19**, 9–23.

Mehmke, R. (1897) Zum Gesetz der Elastichen Dehnungen, *Z. Math. Phys.*, Leipzig, **42**, 327–338.

Miklowitz, J. (1980) *The Theory of Elastic Waves and Waveguides*, North-Holland, Amsterdam et al., 2nd ed.

Mindlin, R.D. and Eshel, N.N. (1968) On First Strain-Gradient Theories in Linear Elasticity, *Int. J. Solids Structures*, **4**, 109–124.

Miropolski, Y.Z. (1981) *Dynamics of Internal Gravitational Waves in the Ocean*, Gidrometeoizdat, Moscow (in Russian).

Miura, R.M. (1976) The Korteweg-de Vries Equation: a Survey of Results, *SIAM Review*, **18**, 412–459.

Miura, R.M. (1982) Accurate Computation of the Stable Solitary Wave for the Fitz Hugh-Nagumo Equations, *J. Math. Biol.*, **13**, 247–269.

Moon, F.C. (1987) *Chaotic Vibrations*, J.Wiley, New York et al.

Muir, T.G. and Carstensen, E.L. (1980) Prediction of Non-Linear Acoustic Effects

of Biomedical Frequencies and Intensities, *Ultrasound Med. Biol.*, **6**, 345–357.
Mura, T. (1993) *Micromechanics of Defects in Solids*, Kluwer Acad. Publ., Dordrecht et al., 2nd ed.
Müller, I. (1987) Kinetic Theory and Extended Thermodynamics, in I.Müller and T.Ruggeri (Eds.), *ISIMM Symp. on Kinetic Theory and Extended Thermodynamics*, Pitagora Editrice, Bologna, 245–258.
Nagumo, J., Arimoto, S. and Yoshizawa, S. (1962) An Active Pulse Transmission Line Simulating Nerve Axon, *Proc. IRE*, **50**, 2061–2070.
Narasimhan, M.N.L. (1993) *Principles of Continuum Mechanics*, J.Wiley, New York.
Nariboli, G.A. and Sedov, A. (1970) Burgers-Korteweg-de Vries Equation for Viscoelastic Rods and Plates, *J. Math. Anal. Appl.*, **32**, 661–677.
Nash, S. (1985) Science and Complexity, *Proc. of an Interdisciplinary IBM Conference*, London, Science Reviews, Ltd., Northwood.
Naugolnykh, K.A. and Ostrovsky, L.A. (1990) *Non-Linear Wave Processes in Acoustics*, Nauka, Moscow (in Russian).
Naumkin, P.J. and Shishmarov, I.A. (1994) *Nonlinear Nonlocal Equations in the Theory of Waves*, Am. Math. Soc., Providence, Rhode Island.
Nazarov, V.E., Ostrovsky, L.A., Soustova, I.A. and Sutin, A.M. (1988) Nonlinear Acoustics of Microinhomogeneous Media, *PEPI*, **50**, 65–73.
New Encyclopedia Britannica (1991) Chicago et al., Vol. 2.
Newell, A.C. (1985) Solitons in Mathematics and Physics, *SIAM*, New York.
Nicolis, J.S. (1986) *Dynamics of Hierarchical Systems*, Springer, Berlin et al.
Nicolis, G. and Prigogine, I. (1989) *Exploring Complexity: an Introduction*, Freeman, New York.
Nikolaev, A.V. (1989) Scattering and Dissipation of Seismic Waves in the Presence of Nonlinearity, *PAGEOPH*, **131**, 687–702.
Nikolajevskij, V.N. (1990) *Mechanics of Porous and Fractured Media*, World Sci. Publ. Co, Singapore et al.
Nimmo, J.C. and Crighton, D.G. (1986) Geometrical and Diffusive Effects in Nonlinear Acoustic Propagation over Long Ranges, *Phil. Trans. Roy. Soc. London*, **A320**, 1–35.
Normand, C., Pomeau, Y. and Velarde, M.G. (1977) Convective Instability: A Physicist's Approach, *Rev. of Modern Phys.*, **49**, 581–624.
Nunciato, J.W., Walsh, E.K., Schuler, K.W. and Barker, L.M. (1974) Wave Propagation in Non-linear Viscoelastic Solids, in S.Flügge (Ed.), *Enc. of Phys.*, **VIa/4**, Springer, Berlin et al., 1–108.
Oliver, D. (1991) *The Shaggy Streed of Physics. Mathematical Beauty in the Physical World*, Springer, New York et al.
Osborne, A.R. (1993) Behavior of Solitons in Random-Function Solutions of the Periodic Korteweg-de Vries Equation, *Phys. Rev. Lett.*, **71**, 3115–3118.
Oseen, C.W. (1933) The Theory of Liquid Crystals, *Transactions of the Faraday Society*, **29**, 883–899.
Ostoja-Starzewski, M. (1991) Transient Waves in a Class of Random Heterogeneous Media, *Appl. Mech. Rev.*, **44**(11), Part 2, 199–209.
Ostrovski, L.A. (1983) Solitons in Active Media, in U.Nigul and J.Engelbrecht

(Eds.), *Nonlinear Deformation Waves*, Springer, Berlin et al., 30–43.

Ostrovski, L.A. (1989) Nonlinear Properties of Elastic Media with Cylindrical Pores, *Soviet Physics-Acoustics*, **35**, 490–494.

Othmer, H.G. Ed. (1986) *Nonlinear Oscillations in Biology and Chemistry*, Springer, Berlin et al.

Packel, E.W. and Traub, J.F. (1987) Informationbased Complexity, *Nature*, **328**, 29–33.

Parker, D.F. (1994) Nonlinear Surface Acoustic Waves and Waves on Stratified Media, in A.Jeffrey and J.Engelbrecht (Eds.), *Nonlinear Waves in Solids*, Springer, Vienna and New York, 289–347.

Parker, D.F. and Talbot, T.M. (1985) Analysis and Computation for Nonlinear Elastic Surface Waves of Permanent Form, *J. Elasticity*, **15**(4), 389–426.

Peipman, T., Valdek, U. and Engelbrecht, J. (1992) Nonlinear Two-Dimensional Longitudinal and Shear Waves in Solids, *Acustica*, **76**, 84–94.

Peitgen, H.-O. and Richter, P.H. (1986) *The Beauty of Fractals. Images of Complex Dynamical Systems*, Springer, Berlin et al.

Peitgen, H.-O. and Saupe, D. (1988) *The Science of Fractal Images*, Springer, New York et al.

Peitgen, H.-O., Jürgens, H. and Saupe, D. (1992) *Fractals for the Classroom, I—Introduction to Fractals and Chaos*, Springer, New York et al.

Pence, T.J. (1992) On the Mechanical Dissipation of Solutions to the Riemann Problem for Impact Involving a Two-Phase Elastic Material, *Arch. Rat. Mech. and Anal.*, **117**, 1–52.

Perelson, A.S., Goldstein, B., Dembo, M. and Jacquez, J.A. Eds. (1988) *Nonlinearity in Biology and Medicine*, Elsevier, New York.

Peterson, I. (1988) *The Mathematical Tourist*, Freeman, New York.

Petrov, V.A. (1983) Dilaton Model of Thermal Fluctuation Crack Nucleation, *Sov. Phys. Solid State*, **25**(10), 1800–1802.

Pierce, J. (1974) *Almost All about Waves*, MIT Press, Cambridge, Mass.

Prigogine, I. and Stengers, I. (1984) *Order out of Chaos*, Heinemann, London.

Ravasoo, A. (1989) Propagation of Non-Linear Waves in Inhomogeneous Hereditary Media, *Int. J. Non-Linear Mech.*, **24**, 57–64.

Ravasoo, A. (1995) Ultrasonic Nondestructive Evaluation of Inhomogeneous Plane Strain in Elastic Medium, *Res. Nondestr. Eval.*, **7**(1), 55–68.

Rayleigh, Lord (1876) On Waves, *Philos. Mag.*, Ser. 5, **1**, 257–259.

Ricker, N.H. (1977) *Transient Waves in Viscoelastic Media*, Elsevier, Amsterdam.

Riley, D.S. (1990) Long-Wave Morphologies in Directional Solidification, *Appl. Mech. Rev.*, **43**(5), Part 2, 85–88.

Rosenau, P. (1992) Modern and Post-Modern Science. Some Contrasts, *Review (Fernand Braudel Center)*, **15**, 49–89.

Rosenau, P. and Hyman, J.M. (1993) Compactons: Solitons with Finite Wavelength, *Phys. Rev. Lett.*, **70**, 564–567.

Sadovski, M.A. (1985) State and Perspecitves of Scientific Research in Prediction of Earthquakes, *Vestnik USSR Acad. Sci.*, **10**, 26–38 (in Russian).

Salupere, A. (1995) On the Application of the Pseudospectral Method for Solving the Korteweg-de Vries Equation, *Proc. Estonian Acad. Sci., Phys. Math.*, **44**(1), 73–87.

Salupere, A., Maugin, G.A. and Engelbrecht, J. (1994) Korteweg-de Vries Soliton Detection from a Harmonic Input, *Phys. Lett.*, **A192**, 5–8.

Salupere, A., Maugin, G.A., Engelbrecht, J. and Kalda, J. (1996) On the KdV Soliton Formation and Discrete Spectral Analysis, *Wave Motion*, **23**, 49–66.

Samsonov, A.M. (1988) On Existence of Longitudinal Strain Solitons in an Infinite Nonlinearly Elastic Rod, *Sov. Phys.-Doklady*, **33**, 298–300.

Samsonov, A.M. (1994) Nonlinear Strain Waves in Elastic Waveguides, in A.Jeffrey and J.Engelbrecht (Eds.), *Nonlinear Waves in Solids*, Springer, Wien and New York.

Sander, J. and Hutter, K. (1991) On the Development of the Theory of the Solitary Wave: a Historical Essay, *Acta Mech.*, **86**, 111-152.

Schattschneider, D. (1990) *Visions of Symmetry*, Freeman, New York.

Scheid, H. (1993) About Chance, *Universitas*, **35**, 124–133.

Schierwagen, A.K. (1991) Travelling Wave Solutions of a Simple Nerve Conduction Equation for Inhomogeneous Axons, in A.V.Holden, M.Markus and H.Othmer (Eds.), *Nonlinear Waves in Excitable Media*, Manchester University Press, Manchester, 107–114.

Schroeder, M. (1991) *Fractals, Chaos, Power Laws*, W.H.Freeman and Co, New York.

Scott, A.C. (1970) *Active and Nonlinear Wave Propagation in Electronics*, J.Wiley, New York et al.

Scott, A.C. (1975) The Electrophysics of a Nerve Fiber, *Rev. Mod. Phys.*, **47**(2), 487–533.

Scott, A.C., Chu F.Y.E. and Mc Laughlin, D.W. (1973) The Soliton: a New Concept in Applied Science, *Proc. IEEE*, **61**, 1443–1486.

Scott Russel, J. (1844) Report on Waves, *Fourteenth Meeting of the British Association for the Advancement of Science*, John Murray, London, 311–390.

Scott Russel, J. (1865) *The Modern System of Naval Architecture*, Day and Son, London, 1.

Shorter Oxford English Dictionary (1962) Clarendon Press, Oxford.

Schuster, H.G. (1984) *Deterministic Chaos. An Introduction*, Physik-Verlag, Weinheim.

Skudrzyk, E. (1971) *The Foundations of Acoustics*, Springer, New York.

Slepyan, L.I. (1990) Mechanics of Cracks, *Sudostroyenie*, Leningrad, 2nd ed. (in Russian).

Slepyan, L., Krylov, V. and Parnes, R. (1995) Solitary Waves in a Helix, *Proc. Estonian Acad. Sci., Phys. Math.*, **44**(1), 29–39.

Smith, D.R. (1993) *An Introduction to Continuum Mechanics—After Truesdell and Noll*, Kluwer, Dordrecht.

Stewart, I. (1989) *Does God Play Dice? The Mathematics of Chaos*, Blackwell, Oxford.

Stronge, W.J. (1987) The Domino Effect: a Wave of Destabilizing Collisions in a Periodic Array, *Proc. Roy. Soc. London*, **A409**, 199–208.

Stronge, W.J. (1989) Group Theory for the Domino Effect, in J.Engelbrecht (Ed.), *Nonlinear Waves in Active Media*, Springer, Berlin, 76–83.

Sugimoto, N. (1989) Steady Propagation of Buckle in Elastic Pipelines, in M.F.McCarthy and M.A.Hayes (Eds.), *Elastic Wave Propagation*, Elsevier, North-Holland, 265–272.

Sugimoto, N. (1993) "Shock-Free Tunnel" for Future High-Speed Trains, in *Proc. Int. Conf. on Speedup Technology for Railway and Maglev Vehicles, ISME*, Yokohama, 284–292.

Sugimoto, N. and Kakutani, T. (1985) "Generalized Burgers' Equation" for Nonlinear Viscoelastic Waves, *Wave Motion*, **7**(5), 447–458.

Suh, N.P. and Turner, A.P.L. (1975) *Elements of the Mechanical Behavior of Solids*, McGraw-Hill, New York et al.

Tanaka, K. and Motoyama, O. (1974) The Propagation of Longitudinal Wave in a Viscoelastic Circular Bar with Consideration of the Lateral Inertia, *Bull. ISME*, **17**, 1247–1257.

Taniuti, T. and Nishihara, K. (1983) *Nonlinear Waves*, Pitman, London et al. (in Japanese 1977).

Taya, M. and Arsenault, R.J. (1989) *Metal Matrix Composites*, Pergamon Press, Oxford.

Thompson, J.M.T. and Stewart, H.B. (1986) *Nonlinear Dynamics and Chaos*, J.Wiley, Chichester et al.

Toda, M. (1989) *Nonlinear Waves and Solitons*, Kluwer, Dordrecht et al.

Toffler, A. (1984) Preface to Prigogine, I., Stengers, I. *Order out of Chaos*, Heinemann, London.

Tritton, D.J. (1988) *Physical Fluid Dynamics*, Clarendon Press, Oxford, 2nd ed.

Truesdell, C. (1968) *Essays in the History of Science*, Springer, Berlin et al.

Truesdell, C. (1984) *An Idiot's Fugitive Essays on Science*, Springer, New York et al.

Truesdell, C.A. and Toupin, R.A. (1960) The Classical Field Theories, in S.Flügge (Ed.), *Enc. of Phys.*, **III/1**, Springer, Berlin et al., 226–293.

Truesdell, C. and Noll, W. (1965) The Non-Linear Field Theories of Mechanics, in S.Flügge (Ed.), *Enc. of Phys.*, **III/3**, Springer, Berlin et al.

de Vries, G. (1894) *Bijdrage Tot de Kennis der Lange Golven*, Doct. Diss., University of Amsterdam.

Waag, R.C. (1984) A Review of Tissue Characterization from Ultrasonic Scattering, *IEEE Trans. Biomed. Eng.*, **BME 31**, 884–893.

Wachbroit, R. (1994) Normality as a Biological Concept, *Philosophy of Science*, **61**, 579–591.

Wadati, M., Konno, K. and Ichikawa, Y.W. (1979) New Integrable Nonlinear Evolution Equations, *J. Phys. Soc. Japan*, **47**, 1689–1700.

West, B.J. (1985) *An Essay on the Importance of Being Nonlinear*, Springer, Berlin et al.

Whitham, G.B. (1974) *Linear and Nonlinear Waves*, J.Wiley, New York et al.

Zabolotskaya, E.A. (1992) Nonlinear Propagation of Plane and Circular Rayleigh Waves in Isotropic Solids, *J. Acoust. Soc. Am.*, **91**(5), 2569–2575.

Zabusky, N.J. (1969) Nonlinear Lattice Dynamics and Energy Shaving, *J. Phys. Soc. Japan*, **26**, Suppl., 196–202.

Zabusky, N.J. (1981) Computational Synergetics and Mathematical Innovation, *J. Comput. Phys.*, **43**, 195–249.

Zabusky, N.J. and Kruskal, M.D. (1965) Interaction of "Solitons" in a Collisionless Plasma and the Recurrence of Initial States, *Phys. Rev. Lett.*, **15**, 240–243.

Zhurkov, S.N. (1983) Dilaton Mechanism of the Strength of Solids, *Sov. Phys. Solid State*, **25**, 1797–1800.

Index

Kluwer Texts in the Mathematical Sciences

1. A.A. Harms and D.R. Wyman: *Mathematics and Physics of Neutron Radiography.* 1986 ISBN 90-277-2191-2
2. H.A. Mavromatis: *Exercises in Quantum Mechanics.* A Collection of Illustrative Problems and Their Solutions. 1987 ISBN 90-277-2288-9
3. V.I. Kukulin, V.M. Krasnopol'sky and J. Horácek: *Theory of Resonances.* Principles and Applications. 1989 ISBN 90-277-2364-8
4. M. Anderson and Todd Feil: *Lattice-Ordered Groups.* An Introduction. 1988 ISBN 90-277-2643-4
5. J. Avery: *Hyperspherical Harmonics.* Applications in Quantum Theory. 1989 ISBN 0-7923-0165-X
6. H.A. Mavromatis: *Exercises in Quantum Mechanics.* A Collection of Illustrative Problems and Their Solutions. Second Revised Edition. 1992 ISBN 0-7923-1557-X
7. G. Micula and P. Pavel: *Differential and Integral Equations through Practical Problems and Exercises.* 1992 ISBN 0-7923-1890-0
8. W.S. Anglin: *The Queen of Mathematics.* An Introduction to Number Theory. 1995 ISBN 0-7923-3287-3
9. Y.G. Borisovich, N.M. Bliznyakov, T.N. Fomenko and Y.A. Izrailevich: *Introduction to Differential and Algebraic Topology.* 1995 ISBN 0-7923-3499-X
10. J. Schmeelk, D. Takači and A. Takači: *Elementary Analysis through Examples and Exercises.* 1995 ISBN 0-7923-3597-X
11. J.S. Golan: *Foundations of Linear Algebra.* 1995 ISBN 0-7923-3614-3
12. S.S. Kutateladze: *Fundamentals of Functional Analysis.* 1996 ISBN 0-7923-3898-7
13. R. Lavendhomme: *Basic Concepts of Synthetic Differential Geometry.* 1996 ISBN 0-7923-3941-X
14. G.P. Gavrilov and A.A. Sapozhenko: *Problems and Exercises in Discrete Mathematics.* 1996 ISBN 0-7923-4036-1
15. R. Singh and N. Singh Mangat: *Elements of Survey Sampling.* 1996. ISBN 0-7923-4045-0
16. C.D. Ahlbrandt and A.C. Peterson: *Discrete Hamiltonian Systems.* Difference Equations, Continued Fractions, and Riccati Equations. 1996 ISBN 0-7923-4277-1
17. J. Engelbrecht: *Nonlinear Wave Dynamics.* Complexity and Simplicity. 1997 ISBN 0-7923-4508-8

KLUWER ACADEMIC PUBLISHERS – DORDRECHT / BOSTON / LONDON

MIX
Papier aus verantwortungsvollen Quellen
Paper from responsible sources
FSC® C105338

If you have any concerns about our products,
you can contact us on
ProductSafety@springernature.com

In case Publisher is established outside the EU,
the EU authorized representative is:
Springer Nature Customer Service Center GmbH
Europaplatz 3, 69115 Heidelberg, Germany

Printed by Libri Plureos GmbH
in Hamburg, Germany